STUDENT UNIT GUIDE

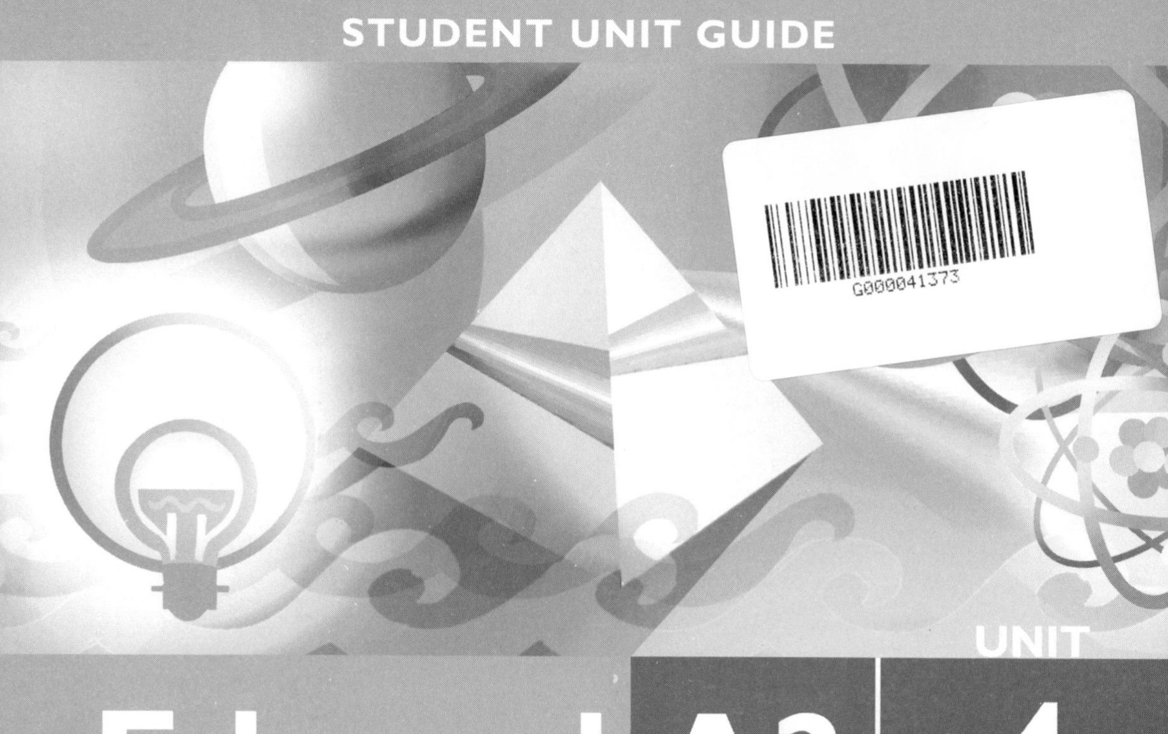

Edexcel A2 UNIT 4

Physics

Physics on the Move

Mike Benn

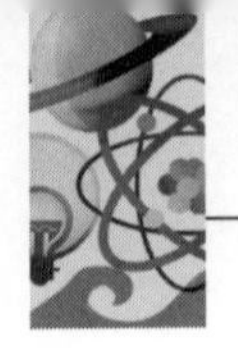

Philip Allan Updates, an imprint of Hodder Education, an Hachette UK company, Market Place, Deddington, Oxfordshire OX15 0SE

Orders
Bookpoint Ltd, 130 Milton Park, Abingdon, Oxfordshire OX14 4SB
tel: 01235 827720
fax: 01235 400454
e-mail: uk.orders@bookpoint.co.uk
Lines are open 9.00 a.m.–5.00 p.m., Monday to Saturday, with a 24-hour message answering service. You can also order through the Philip Allan Updates website: www.philipallan.co.uk

ISBN 978-0-340-94827-9

First printed 2010
Impression number 5 4 3 2 1
Year 2015 2014 2013 2012 2011 2010

This guide has been written specifically to support students preparing for the Edexcel A2 Physics Unit 4 examination. The content has been neither approved nor endorsed by Edexcel and remains the sole responsibility of the author.

Typeset by Pantek Arts Ltd, Maidstone, Kent.
Printed by MPG Books, Bodmin

Contents

Introduction

Content Guidance

Questions and Answers

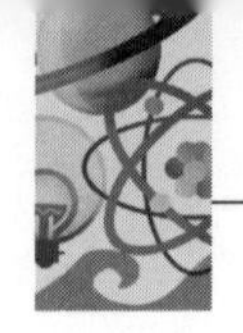

Introduction

About this guide

This guide is one of a series covering the Edexcel specification for AS and A2 physics. It offers advice for the effective study of **Unit 4: Physics on the Move**. Its aim is to help you *understand* the physics — it is not intended as a shopping list that enables you to cram for the examination. The guide has three sections:

- **Introduction** — this gives brief guidance on approaches and techniques to ensure you answer the examination questions in the best way that you can.
- **Content Guidance** — this section is not intended to be a detailed textbook. It offers guidance on the main areas of the content of Unit 4, with an emphasis on worked examples. The examples illustrate the types of question that you are likely to come across in the examination.
- **Questions and Answers** — this comprises two unit tests, presented in a format close to that of an actual Edexcel examination and using questions similar to those in recent past papers, to give the widest possible coverage of the unit content. Answers are provided; in some cases, distinction is made between responses that might have been written by an A-grade candidate and those typical of a C-grade candidate. Common errors made by candidates are also highlighted so that you, hopefully, do not make the same mistakes.

Understanding physics requires time and effort. No one suggests that physics is an easy subject, but even students who find it difficult can overcome their problems by the proper investment of time.

A deep understanding of physics can develop only with experience, which means time spent thinking about physics, working with it and solving problems. This book provides you with a platform for doing this. If you try all the worked examples and the unit tests *before* looking at the answers, you will begin to think for yourself and develop the necessary techniques for answering examination questions effectively. In addition, you will need to *learn* the all basic formulae, definitions and experiments. Thus prepared, you will be able to approach the examination with confidence.

The specification

The specification outlines the physics that will be examined in the unit tests and describes the format of those tests. This is not necessarily the same as what teachers might choose to teach or what you might choose to learn.

The purpose of this book is to help you with Unit Test 4, following the specification; but don't forget that what you are doing is learning *physics*, therefore you should also consult a standard textbook for more information. Physics is a logical discipline and, to truly understand it, you need to have a feeling for how the concepts you are dealing with fit into the broader picture of the subject.

The specification can be obtained from Edexcel, either as a printed document or from the web at **www.edexcel.com**.

The unit test

Unit Test 4 is a written paper of duration 1 hour and 35 minutes. It carries a total of 80 marks, accounting for 40% of the A2 or 20% of the total A-level marks. There are ten objective (multiple-choice) questions, each worth a single mark, followed by short and long questions worth between 3 and 16 marks each. The questions will assume that Units 1 and 2 have been studied, but will not examine the content of these units again in detail. For instance, knowledge of forces and Newton's laws will be expected in a question on circular motion.

The test will examine the objectives AO1 (knowledge and understanding of science and of how science works), AO2 (application of knowledge and understanding) and AO3 (how science works). Compared with the AS papers, there is a slight shift of emphasis towards AO2. Although the questions are structured to some extent, they are less so than those you encountered in the AS units. In particular, quantities are often not given in base units so that, for example, you may need to convert measurements stated in nanometres and millimetres to their equivalent values in metres before substituting into the appropriate equation.

Command terms

Examiners use certain words that require you to respond in a particular way. You must be able to distinguish between these terms and understand exactly what each requires you to do. Some frequently used commands are shown below.

- **State** — the answer should be a brief sentence giving the essential facts; no explanation is required (nor should you give one).
- **Define** — you can use a *word equation*; if you use *symbols*, it is essential to state clearly what each symbol represents.
- **List** — simply give a series of words or terms; there is no need to write sentences.
- **Outline** — this word is often used when asking you to give a brief description of a process; a logical series of bullet points or phrases will suffice.
- **Describe** — for an experiment, a diagram is essential; then state the main points concisely (bullet points can be used).
- **Draw** — diagrams should be drawn in section, neatly and *fully labelled* with all measurements clearly shown; but don't waste time — remember that this is not an art exam.
- **Sketch** — usually a graph is called for, but graph paper is not necessary (although a grid is sometimes provided); axes must be labelled and include a scale if numerical data are given; the origin should be shown if appropriate, and the general shape of the expected line or curve should be drawn clearly.

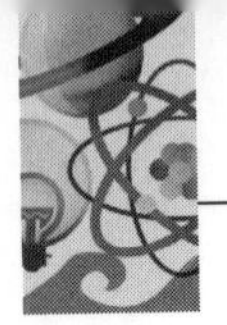

- **Explain** — use correct physics terminology and principles; the amount of detail in your answer should reflect the number of marks available.
- **Show that** — usually a value is provided (to enable you to proceed with the next part of the question) and you have to demonstrate how this value can be obtained; you should show all your working and state your result to more significant figures than the given value contains (to prove that you have actually done the calculations).
- **Calculate** — show all your working and include *units* at every stage; the number of significant figures in your answer should reflect the given data, but you should keep each stage with more significant figures in your calculator to prevent excessive rounding.
- **Determine** — you will probably have to extract some data, often from a graph, in order to perform a calculation.
- **Estimate** — this means doing a calculation in which you have to make a sensible assumption, possibly about the value of one of the quantities; think — does your assumption lead to a reasonable answer?
- **Suggest** — there is often no single correct answer; credit is given for sensible reasoning based on correct physics.
- **Discuss** — you need to sustain an argument, giving evidence for and against, based on your knowledge of physics and possibly using appropriate data to justify your answer.

You should pay particular attention to diagrams, graph-sketching and calculations. Candidates often lose marks by failing to label diagrams properly, by not giving essential numerical data on sketch graphs, and by not showing all their working or by omitting units in calculations.

Revision

This purpose of this introduction is not to provide you with an in-depth guide to revision techniques — you can find many books on study skills if you feel you need more help in preparing for examinations. There are, however, some points worth mentioning that will help you when you are revising the physics A-level material:

- Familiarise yourself with what you need to know — ask your teacher and look through the specification.
- Make sure you have a good set of notes — you can't revise properly from a textbook.
- Learn all the equations indicated in the specification and be familiar with the formulae that will be provided in the examination (at the end of each question paper) so that you can find them quickly and use them correctly.
- Make sure that you learn definitions thoroughly and in detail, e.g. the principle of conservation of linear momentum requires that in a *system of interacting bodies* the total momentum is conserved, provided that *no resultant external force acts on the system*.
- Be able to describe (with diagrams) the basic experiments referred to in the specification.
- Make revision active by writing out equations and definitions, drawing diagrams, describing experiments and performing lots of calculations. Remember — practice makes perfect!

Content Guidance

This section is a guide to the content of **Unit 4: Physics on the Move**, which contains three main topics.

Further mechanics

- Momentum — $p = mv$; principle of conservation of momentum and its application to problems in one and two dimensions; Newton's second law in terms of rate of change in momentum; impulse
- Conservation of energy — elastic and inelastic collisions; derivation and use of $E_k = p^2/2m$
- Circular motion — angular displacement in degrees and radians; angular velocity and use of $v = \omega r$ and $T = 2\pi/\omega$; centripetal acceleration and force $F = mv^2/r = m\omega r^2$

Electric and magnetic fields

- Electric fields — field strength $E = F/Q$; using lines of force to represent uniform and radial fields; Coulomb's law $F = kQ_1Q_2/r^2$ and radial electric field $E = kQ/r^2$ due to a point charge; $E = V/d$ between a pair of parallel charged plates
- Capacitance — $C = Q/V$; energy stored in a capacitor, $W = \frac{1}{2}QV$, and related expressions $W = \frac{1}{2}CV^2 = \frac{1}{2}Q^2/C$; use of $Q = Q_0e^{-t/RC}$ and related formulae for the charge and discharge of capacitors
- Magnetic fields — flux density, flux, flux linkage; use of $F = BIl\sin\theta$, $F = Bqv\sin\theta$ and Fleming's left-hand rule
- Electromagnetic induction — factors affecting the induced emf in a coil when there is a change in magnetic flux linkage; Faraday's and Lenz's laws and the use of $\varepsilon = -\frac{d(N\Phi)}{dt}$

Particle physics

- Atomic structure — proton and nucleon numbers; the nuclear atom; alpha particle scattering experiment
- Particle accelerators and detectors — role of electric and magnetic fields; electron guns, linacs, cyclotrons; use of $r = p/BQ$ to interpret particle tracks
- Particle interactions — conservation of charge, energy, and momentum; $\Delta E = c^2\Delta m$; non-SI units MeV, GeV and u, and their mass or energy equivalents
- Subatomic particles — the quark–lepton model; baryons and mesons; standard nuclear notation; wave–particle duality and de Broglie's wavelength $\lambda = h/p$

Quantity algebra

In the worked examples and answers to sample test questions, quantity algebra is used throughout. This involves putting into an equation the units for each quantity. For example, to calculate the speed of an electron given that its effective wavelength is 0.330 nm, we would write:

$$p = mv = \frac{h}{\lambda} \Rightarrow v = \frac{h}{m\lambda} = \frac{6.63 \times 10^{-34}\,\text{J s}}{9.11 \times 10^{-31}\,\text{kg} \times 0.330 \times 10^{-9}\,\text{m}} = 2.2 \times 10^{6}\,\text{m s}^{-1}$$

Although Edexcel does not require you to use quantity algebra, you are strongly advised to do so because of the following advantages:

- It acts as a reminder to substitute consistent units — in the above example, 0.330 nm is written as 0.330×10^{-9} m.
- It allows you to check that the units of the answer are correct — in this example, J is equivalent to N m, which is the same as kg m s^{-2} × m, so the units in the numerator and denominator partially cancel, leaving the correct unit of speed, m s^{-1}.

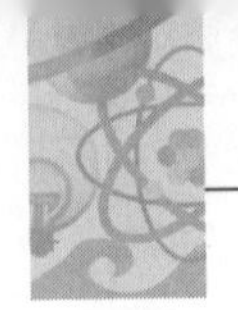

Further mechanics

From Unit 1 you should be familiar with:

- velocity and acceleration as vectors with magnitude and direction
- Newton's laws of motion relating to bodies of fixed mass

This unit involves the study of:

- momentum and the principle of conservation of momentum
- Newton's second law in terms of changes in momentum
- circular motion of bodies moving with constant speed

The topics of momentum and circular motion depend heavily on knowledge and understanding of the basic mechanics covered in Unit 1. You may, therefore, find it helpful to look back over that unit to re-familiarise yourself with its contents.

Momentum

Definition

- momentum = mass × velocity

  $p = mv$ units: kg m s^{-1}

Momentum is a **vector** quantity, with both magnitude and direction.

Worked example

(a) Calculate the (magnitude of the) momentum of:

(i) a ball of mass 250 g thrown at speed $8.0\,\text{m s}^{-1}$

(ii) an electron moving at 1% of the speed of light

(b) Estimate the (magnitude of the) momentum of:

(i) a charging elephant

(ii) a scurrying mouse

Answer

(a) (i) $p = mv = 0.250\,\text{kg} \times 8.0\,\text{m s}^{-1} = 2.0\,\text{kg m s}^{-1}$

(ii) $p = mv = 9.11 \times 10^{-31}\,\text{kg} \times 3.0 \times 10^{8}\,\text{m s}^{-1} = 2.7 \times 10^{-22}\,\text{kg m s}^{-1}$

Tip The electron mass and the speed of light are given in the data sheet which is included in the specification (as Appendix 7) and printed at the end of each examination paper.

(b) (i) $5000\,\text{kg} \times 8\,\text{m s}^{-1} = 4 \times 10^{4}\,\text{kg m s}^{-1}$

(ii) $0.02\,\text{kg} \times 1\,\text{m s}^{-1} = 0.02\,\text{kg m s}^{-1}$

Tip For estimates, any reasonable values are acceptable. Here it is assumed that an elephant has a mass of several tonnes and can charge with a speed similar to that of a sprinting athlete, whereas a mouse has a mass of less than 100 g and can run at a speed of only a few metres per second.

The above examples illustrate the concept of momentum. The charging elephant has a large momentum, so stopping it or changing the direction of its motion would be difficult. On the other hand, it would require little effort to alter the state of the mouse's motion.

It is important to remember that momentum is a vector quantity and, as such, requires a direction in addition to the magnitude. Often, as in the examples above, the direction is implied; however, strictly speaking the direction should always be specified — for instance, by stating that the momentum of the elephant is $4 \times 10^4\,\text{kg m s}^{-1}$ due south. Information about the direction is very significant when studying problems in which moving bodies collide and changes in momentum take place.

Worked example

A rubber ball of mass 0.15 kg is dropped and attains a speed of $5.0\,\text{m s}^{-1}$ the instant it strikes the floor. It rebounds upwards with an initial speed of $4.0\,\text{m s}^{-1}$. Calculate:

(a) the momentum of the ball immediately before and after the rebound

(b) the change in momentum of the ball during the time it is in contact with the floor

Answer

(a) Just before hitting the ground,

$p_1 = 0.15\,\text{kg} \times 5.0\,\text{m s}^{-1} = 0.75\,\text{kg m s}^{-1}$ downwards

After the rebound,

$p_2 = 0.15\,\text{kg} \times 4.0\,\text{m s}^{-1} = 0.60\,\text{kg m s}^{-1}$ upwards

(b) If we assign a positive value to the upward velocity, then the downward velocity takes a negative sign.

So change in momentum of the ball $= p_2 - p_1$

$= (0.60\,\text{kg m s}^{-1}) - (-0.75\,\text{kg m s}^{-1})$

$= 1.35\,\text{kg m s}^{-1}$

Principle of conservation of linear momentum

In any system of interacting bodies the total momentum is conserved, provided that no resultant external force acts on the system.

It is important that you fully understand the meaning of the expression 'system of interacting bodies' in the statement. It is evident that if two snooker balls collide and one of the balls is slowed down while the other speeds up, the momentum of each of the balls will change. However, the *vector sum* of the momentums of the balls after the collision will be the same as that immediately before the interaction.

The 'system' for the bouncing ball in the previous example is more difficult to imagine. The Earth is the second body in the system, but it is so massive compared to the ball that the change in its momentum due to the interaction will be imperceptible; nonetheless, the principle of conservation of momentum still applies.

Worked example

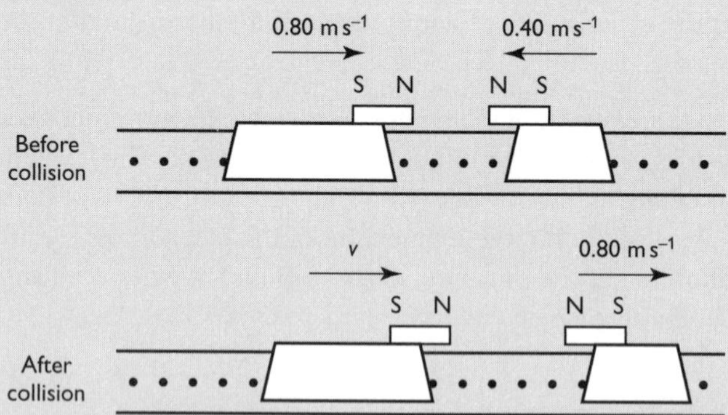

Two gliders moving towards each other on an air track collide and then bounce apart, as shown in the diagram. Initially the 400 g glider travels from left to right with a velocity of $0.80\,\text{m s}^{-1}$ and the 200 g glider moves in the opposite direction at $0.40\,\text{m s}^{-1}$. After the collision the smaller glider changes direction and moves back with a velocity of $0.80\,\text{m s}^{-1}$.

(a) Calculate the velocity of the larger glider after the collision.

(b) If the gliders stuck together on impact, calculate the velocity of the combination.

Answer

(a) Let the rightward direction correspond to positive values, and let v denote the velocity of the larger glider after the collision. Then, for this system, initial momentum $= 0.400\,\text{kg} \times 0.80\,\text{m s}^{-1} + 0.200\,\text{kg} \times (-0.40\,\text{m s}^{-1}) = 0.240\,\text{kg m s}^{-1}$
final momentum $= 0.400\,\text{kg} \times v + 0.200\,\text{kg} \times 0.80\,\text{m s}^{-1}$
Conservation of momentum gives
$0.240\,\text{kg m s}^{-1} = 0.400\,\text{kg} \times v + 0.200\,\text{kg} \times 0.80\,\text{m s}^{-1}$
so $v = +0.20\,\text{m s}^{-1}$
Thus the larger glider moves with a velocity of $0.20\,\text{m s}^{-1}$ to the right.

(b) $0.240\,\text{kg m s}^{-1} = (0.400\,\text{kg} + 0.200\,\text{kg}) \times v_{comb}$
$v_{comb} = +0.40\,\text{m s}^{-1}$
The combination of two gliders moves with a velocity of $0.40\,\text{m s}^{-1}$ to the right.

For oblique collisions in which the masses approach or separate along different lines, the principle of conservation of momentum can be applied to the *components* of the momentum in any direction. (Components of vectors are covered in Unit 1.)

Worked example

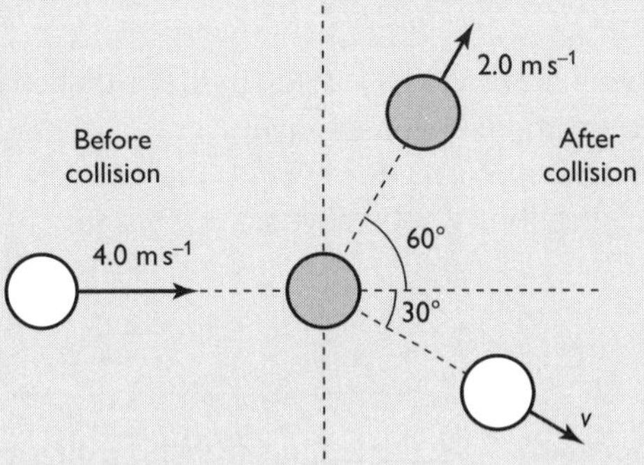

A snooker ball travelling at $4.0\,\mathrm{m\,s^{-1}}$ strikes a stationary ball of identical mass, causing it to move with speed $2.0\,\mathrm{m\,s^{-1}}$ at an angle of 60° to the direction of the incident ball. Calculate the velocity of the incident ball after the collision, given that it is deviated by 30° from its original path.

Answer

Let m kg be the mass of each ball.

Taking components along the direction in which the snooker ball travels before the collision:

initial momentum = $m\,\mathrm{kg} \times 4.0\,\mathrm{m\,s^{-1}} + m\,\mathrm{kg} \times 0\,\mathrm{m\,s^{-1}} = 4.0m\,\mathrm{kg\,m\,s^{-1}}$

final momentum = $m\,\mathrm{kg} \times v\cos 30° + m\,\mathrm{kg} \times 2.0\,\mathrm{m\,s^{-1}}\cos 60°$

As momentum is conserved, we have

$4.0m\,\mathrm{kg\,m\,s^{-1}} = mv\cos 30°\,\mathrm{kg} + 2.0m\cos 60°\,\mathrm{kg\,m\,s^{-1}}$

The masses m cancel, giving $v = 3.5\,\mathrm{m\,s^{-1}}$

Note: The same result can be achieved by taking components of the momentum at right angles to the direction of the incident ball, so that the total momentum is zero. Trying this would be a worthwhile exercise.

Newton's second law of motion

In Unit 1 you learned a version of Newton's second law of motion that applies only to the special case where a resultant force acts upon a body of fixed mass and causes the body to accelerate. The law can be stated more generally in terms of the change in momentum that ensues when a resultant force is applied to a system:

The rate of change in momentum is directly proportional to the resultant applied force.

In symbols:

$$\sum F = \frac{\Delta p}{\Delta t} \qquad \text{(if consistent SI units are used)}$$

This statement of Newton's second law encompasses situations in which the mass is indeterminate (e.g. interactions involving neutrinos) or continually changing (e.g. a stream of water striking a surface). However, in the unit test, Newton's second law will be applied only in situations where the mass is constant.

For a fixed mass:

$$\sum F = \frac{\Delta(mv)}{\Delta t} = m\frac{\Delta v}{\Delta t} = ma$$

The above calculation confirms the validity of the equation used to represent Newton's second law in Unit 1 when the force is applied to a fixed mass.

Impulse

An impulse refers to a change in momentum, usually a change that takes place over a short interval of time:

impulse = change in momentum = $mv - mu$

where u is the velocity at the beginning and v the velocity at the end of the time interval.

Rearranging the expression for Newton's second law in the fixed-mass case, we get:

$$\left(\sum F\right)(\Delta t) = \Delta p = mv - mu$$

It follows that an impulse can also be defined as the product $(\sum F)\Delta t$, i.e. resultant force × time duration, and has the unit N s (newton seconds).

As an exercise, show that the unit of impulse, N s, is consistent with the unit for change in momentum, kg m s^{-1}.

If a graph of force against time is drawn for a short interaction such as a golf ball being struck by the face of a golf club, the impulse can be estimated from the area under the curve:

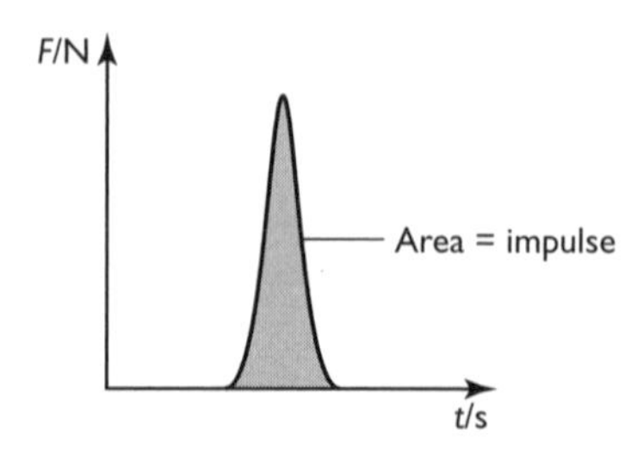

Worked example

In an experiment to investigate Newton's second law, a glider is pulled along an air track by a falling mass, as shown in the diagram.

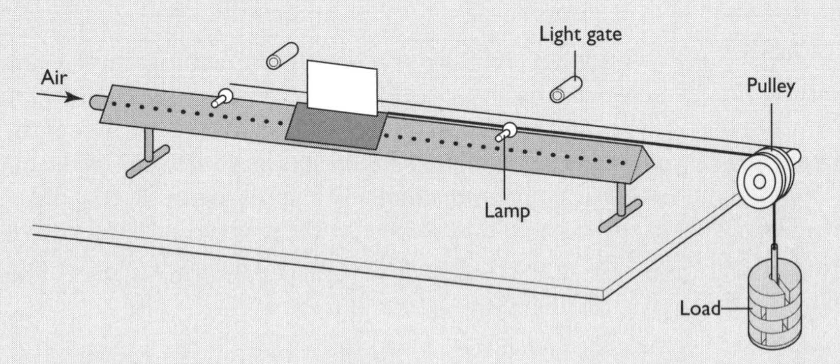

The time taken for the interrupter card to pass through each light gate is recorded, together with the time taken for the glider to travel between the gates. The results of one set of measurements are:

Mass of glider	400 g
Length of interrupter card	20.0 cm
Time to cross first gate	0.316 s
Time to cross second gate	0.224 s
Time between gates	0.524 s

Use these results to calculate:

(a) the initial momentum of the glider

(b) the final momentum of the glider

(c) the resultant force acting on the glider

The mass on the load is 22.0 g.

(d) Comment on how this compares with the resultant force on the glider calculated in part (c).

Answer

(a) Initial velocity $= \dfrac{0.200\,\text{m}}{0.316\,\text{s}} = 0.633\,\text{m}\,\text{s}^{-1}$

Initial momentum $= 0.400\,\text{kg} \times 0.633\,\text{m}\,\text{s}^{-1} = 0.253\,\text{kg}\,\text{m}\,\text{s}^{-1}$

(b) Final velocity $= \dfrac{0.200\,\text{m}}{0.224\,\text{s}} = 0.893\,\text{m}\,\text{s}^{-1}$

Final momentum $= 0.400\,\text{kg} \times 0.893\,\text{m}\,\text{s}^{-1} = 0.357\,\text{kg}\,\text{m}\,\text{s}^{-1}$

(c) Resultant force $= \dfrac{\text{change in momentum}}{\text{time}} = \dfrac{(0.357 - 0.253)\,\text{kg}\,\text{m}\,\text{s}^{-1}}{0.524\,\text{s}} = 0.198\,\text{N}$

(d) The force pulling the glider is the weight of the load, i.e. $0.0220\,\text{kg} \times 9.81\,\text{m}\,\text{s}^{-2} = 0.216\,\text{N}$

It follows that there must be a resistive force of 0.018 N acting on the glider, so that the resultant force is 0.216 N – 0.018 N = 0.198 N as found in part (c).

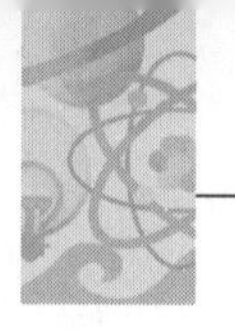

Conservation of energy

Energy cannot be created or destroyed, although it can be transferred from one form to another. This principle and the conservation of momentum together form the cornerstone of physics at all levels.

In collisions both conservation principles must apply, but while the momentum of the system after the interaction is always exactly the same as it was before the interaction, the energy can be changed into other forms. For example, if two identical cars moving in opposite directions at the same speed collide, they will both stop on impact, thus appearing to have lost both momentum and kinetic energy. In fact, the initial momentum of the system was zero because the cars were travelling in opposite directions, and upon impact the kinetic energy of the cars is converted to other forms of energy, predominantly thermal energy.

Collisions are usually classified in terms of the *kinetic* energy in the system before and after the interaction.

- In **elastic collisions**, all of the kinetic energy in the system is conserved.
- In **inelastic collisions**, some or all of the kinetic energy is transferred to other forms of energy.

In everyday life, most collisions are inelastic; however, for interactions involving subatomic particles, elastic collisions are not uncommon. It is also possible for there to be an increase in kinetic energy in a system — for example, in explosions where some chemical energy is converted to kinetic energy of the exploded fragments, and in nuclear radiation and nuclear fission where a mass defect is transferred as extra kinetic energy.

Worked example

(a) Revisit the worked example involving collision of a 400 g glider and a 200 g glider, and use the data to determine the kinetic energy of the gliders before and after the two given types of collision. Hence state whether each collision is elastic or inelastic.

(b) A proton of mass 1.7×10^{-27} kg, moving at $2.0 \times 10^{6}\,\text{m s}^{-1}$, collides with a stationary alpha particle of mass 6.8×10^{-27} kg and rebounds directly backward. The alpha particle moves along the same line with a velocity of $8.0 \times 10^{5}\,\text{m s}^{-1}$. Show that the collision is elastic.

Answer

(a) In the scenario where the gliders rebound:

$$\text{initial kinetic energy} = \frac{1}{2} \times 0.400\,\text{kg} \times (0.80\,\text{m s}^{-1})^2 + \frac{1}{2} \times 0.200\,\text{kg} \times (0.40\,\text{m s}^{-1})^2 = 0.144\,\text{J}$$

$$\text{final kinetic energy} = \frac{1}{2} \times 0.400\,\text{kg} \times (0.20\,\text{m s}^{-1})^2 + \frac{1}{2} \times 0.200\,\text{kg} \times (0.80\,\text{m s}^{-1})^2 = 0.072\,\text{J}$$

The collision is inelastic as 0.072 J of kinetic energy has been transferred to other forms.

In the scenario where the gliders stick together:

initial kinetic energy = 0.144 J (as in the rebound case)

final kinetic energy = $\frac{1}{2} \times (0.400 + 0.200)\,\text{kg} \times (0.40\,\text{m s}^{-1})^2 = 0.048\,\text{J}$

The collision is inelastic as 0.096 J of kinetic energy has been transferred to other forms.

(b) First, we use the law of conservation of momentum to determine the velocity of the rebounding proton.

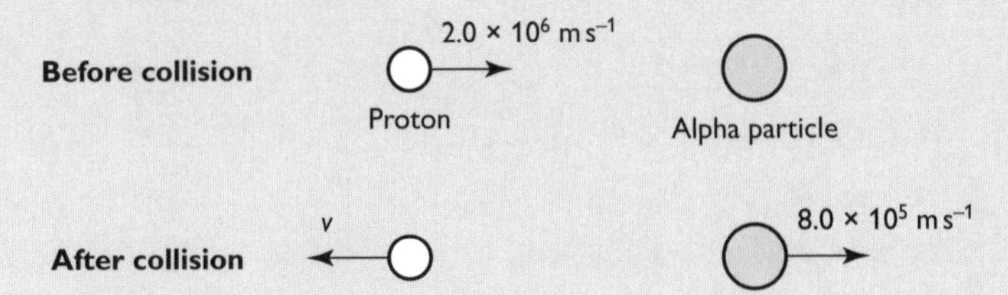

Momentum before the collision

$= 1.7 \times 10^{-27}\,\text{kg} \times 2.0 \times 10^{6}\,\text{m s}^{-1} = 3.4 \times 10^{-21}\,\text{kg m s}^{-1}$

Momentum after the collision

$= 1.7 \times 10^{-27}\,\text{kg} \times v + 6.8 \times 10^{-27}\,\text{kg} \times 8.0 \times 10^{5}\,\text{m s}^{-1}$

Conservation of momentum gives

$1.7 \times 10^{-27}\,\text{kg} \times v + 6.8 \times 8.0 \times 10^{-22}\,\text{kg m s}^{-1} = 3.4 \times 10^{-21}\,\text{kg m s}^{-1}$

so $v = -1.2 \times 10^{6}\,\text{m s}^{-1}$

Initial kinetic energy $= \frac{1}{2} \times (1.7 \times 10^{-27}\,\text{kg}) \times (2.0 \times 10^{6}\,\text{m s}^{-1})^2 = 3.4 \times 10^{-15}\,\text{J}$

Final kinetic energy

$= \frac{1}{2} \times (1.7 \times 10^{-27}\,\text{kg}) \times (-1.2 \times 10^{6}\,\text{m s}^{-1})^2 + \frac{1}{2} \times (6.8 \times 10^{-27}\,\text{kg}) \times (8.0 \times 10^{5}\,\text{m s}^{-1})^2$

$= 3.4 \times 10^{-15}\,\text{J}$

The kinetic energy is the same before and after the collision. Therefore the collision is elastic.

Note: Momentum is a vector quantity, so the negative sign of v indicates that the proton travels in the opposite direction after the collision. Kinetic energy, on the other hand, is a scalar quantity and the direction is irrelevant.

Kinetic energy and momentum

It is often useful, particularly when studying collisions between subatomic particles, to convert momentum to kinetic energy and vice versa. We can derive an expression linking the two quantities $p = mv$ and $E_k = \frac{1}{2}mv^2$.

$$p = mv \Rightarrow p^2 = m^2v^2$$

$$\text{so } \frac{p^2}{2m} = \frac{m^2v^2}{2m} = \frac{1}{2}mv^2$$

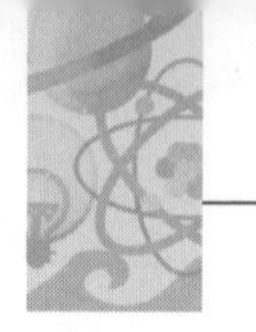

and hence:

$$E_k = \frac{p^2}{2m}$$

Worked example

(a) Calculate the kinetic energy of an electron moving with a momentum of 2.0×10^{-23} kg m s^{-1}.

(b) Calculate the momentum of a rifle bullet of mass 5.0 g fired with kinetic energy 400 J.

Answer

(a) $E_k = \frac{p^2}{2m} = \frac{(2.0 \times 10^{-23}\,\text{kg m s}^{-1})^2}{2 \times (9.1 \times 10^{-31}\,\text{kg})} = 2.2 \times 10^{-16}\,\text{J}$

Note: In particle physics, speeds often approach the speed of light ($c = 3.0 \times 10^8$ m s^{-1}). Be aware, however, that the $E_k = \frac{p^2}{2m}$ equation can be applied only in *non-relativistic* situations, where the speed of the particles is less than about 10% of c. The mass of an electron used here is provided on the data sheet.

(b) $p = \sqrt{2mE_k} = \sqrt{2 \times 5.0 \times 10^{-3}\,\text{kg} \times 400\,\text{J}} = 2.0\,\text{kg m s}^{-1}$

Circular motion

Consider an object moving in a circle of radius r.

Definitions

- **radian** measure of **angular displacement**

  $$\theta = \frac{\text{arc length}}{\text{radius}} = \frac{\Delta s}{r}$$

  2π radians = 360°   1 radian ≈ 57°

- **angular speed**

  $$\omega = \frac{\Delta\theta}{\Delta t} = \frac{\Delta s / r}{\Delta t} = \frac{\Delta s / \Delta t}{r} = \frac{v}{r}$$

  where v is the linear speed of the object (and so $v = \omega r$)

- **period**

  $$T = \frac{2\pi}{\omega}$$

- **frequency**

  $$f = \frac{1}{T} = \frac{\omega}{2\pi}$$

Worked example

(a) Show that the angular speed at which the Moon orbits the Earth is approximately $3 \times 10^{-6}\,\text{rad}\,\text{s}^{-1}$.

(b) Assuming that the Moon has a circular orbit of radius $3.9 \times 10^{8}\,\text{m}$ around the Earth, calculate its linear speed.

Answer

(a) $\omega = \dfrac{2\pi}{T}$ where T, the time taken for the Moon to orbit the Earth once, is 28 days.

Thus $\omega = \dfrac{2\pi\ \text{rad}}{28 \times 24 \times 60 \times 60\ \text{s}} = 2.6 \times 10^{-6}\,\text{rad}\,\text{s}^{-1} \approx 3 \times 10^{-6}\,\text{rad}\,\text{s}^{-1}$

(b) $v = \omega r = 2.6 \times 10^{-6}\,\text{rad}\,\text{s}^{-1} \times 3.9 \times 10^{8}\,\text{m} = 1.0 \times 10^{3}\,\text{m}\,\text{s}^{-1}$

Centripetal acceleration

Consider a body of mass m moving in a circle of radius r at constant speed.

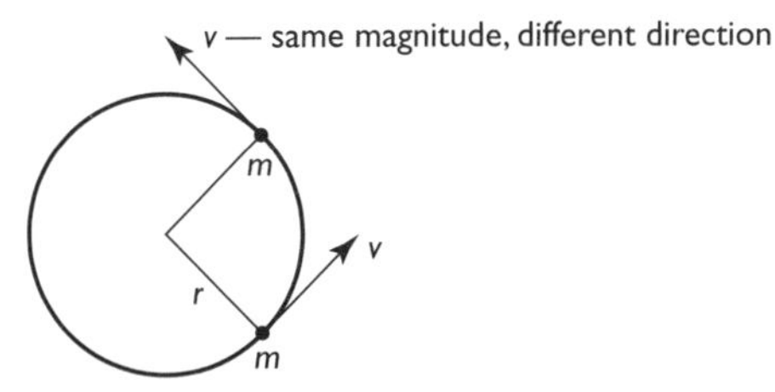

From the diagram we can see that while the magnitude of v is constant, its direction is continuously changing; in other words, although the linear speed is constant, the *velocity*, which is a vector quantity, is always changing.

- The change in velocity means that the mass is accelerating.
- Since the magnitude of v remains unchanged, the acceleration must always be directed at right angles to v, towards the centre of the circle.
- The magnitude of the acceleration is given by: $a = \dfrac{v^2}{r}$

This acceleration is known as **centripetal acceleration**.

Using the relation $v = \omega r$, centripetal acceleration can also be expressed in terms of the angular speed:

$$a = \omega^2 r$$

By Newton's second law, a *resultant force* is required to produce the centripetal acceleration. This resultant force, called the **centripetal force**, is directed towards the centre of the circle and has a magnitude given by:

$$F = ma = \frac{mv^2}{r} = m\omega^2 r$$

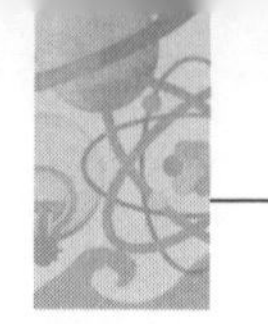

The following diagram shows the forces acting on a water skier.

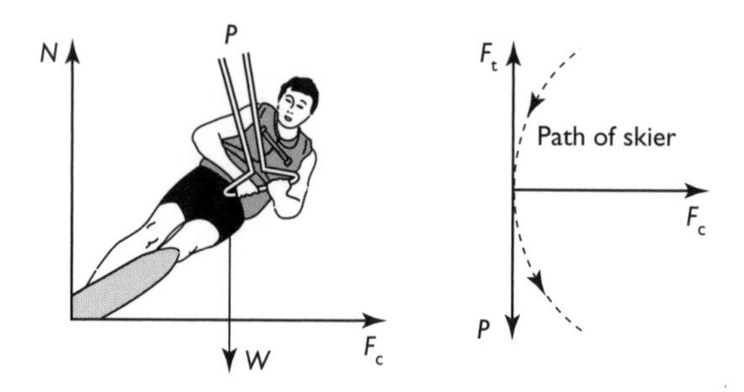

The tangential speed is constant, so, by Newton's first law, the resultant tangential force must be zero. Therefore the forward pull on the rope equals the backward drag of the water on the skis, i.e. $P = F_t$.

If the mass of the skier (plus board) is m, then we have $F_c = \frac{mv^2}{r}$. The centripetal force F_c comes from the sideways push of the water on the skis; so, according to Newton's third law, the sideways push of the skis on the water is $-F_c$.

Worked example

A satellite orbits the Earth once every 87 minutes.

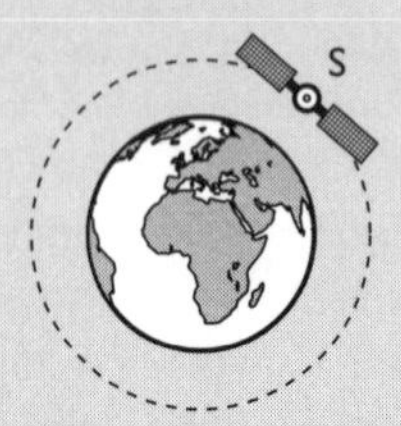

(a) Show that the angular speed of the satellite is approximately 1×10^{-3} rad s^{-1}.

(b) Draw a free-body force diagram for the satellite when it is in the position shown.

(c) With reference to your free-body force diagram, explain why the satellite is accelerating.

(d) The radius of the satellite's orbit is 6500 km. Calculate the magnitude of its acceleration.

Answer

(a) $T = 87 \times 60$ seconds, so:

$$\omega = \frac{2\pi \text{ rad}}{87 \times 60 \text{ s}} = 1.2 \times 10^{-3} \text{ rad s}^{-1} \approx 1 \times 10^{-3} \text{ rad s}^{-1}$$

Tip Don't forget that in a 'show that' question, as evidence that you have actually gone through the calculation, you need to work out the answer to at least one more significant figure than in the given value.

(b)

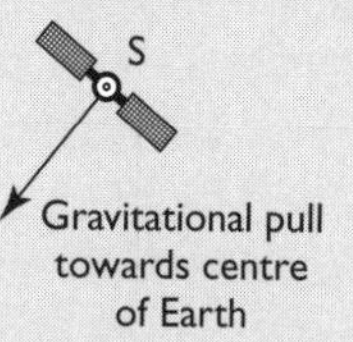

(c) The satellite is acted on by a single resultant force which is directed towards the centre of the Earth; it will therefore accelerate in this direction.

(d) $a = \omega^2 r = (1.2 \times 10^{-3}\,\text{rad}\,\text{s}^{-1})^2 \times (6500 \times 10^3\,\text{m}) = 9.4\,\text{m}\,\text{s}^{-2}$

In circular motion problems you should be aware of the cause of the centripetal force. In the example of the water skier, it is the push of the water against the skis that provides the resultant force; for the satellite, the gravitational pull of the Earth is what keeps it in orbit. Other sources of centripetal force include the tension in a string when a mass attached to the end of the string is whirled in a circle, and the frictional force of the road pushing on the tyres as a car rounds a bend.

Electric and magnetic fields

In physics, a field represents a region in which certain objects will experience a force. In Unit 2 you studied some of the behaviour and properties of electromagnetic radiation, including the associated variation of electric and magnetic fields in waves. In this unit, we look into the separate properties of electric and magnetic fields, and examine how the two are linked in the process of electromagnetic induction.

Electric fields

An electric field is a region in which a charged object will experience a force.

An electric field is generally represented by drawing **field lines** or **lines of force** which show the direction of the force that a very small positive charge would be subject to if it were placed in the field.

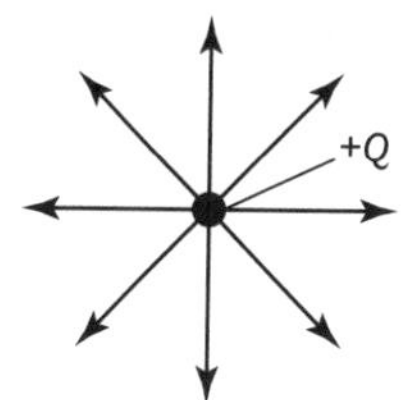

Radial field around a point charge

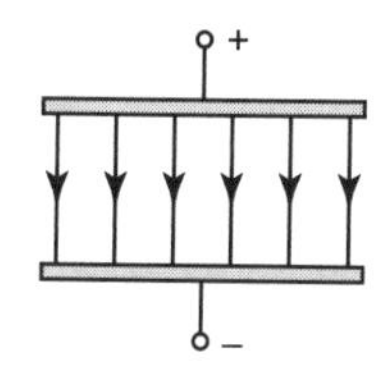

Uniform field between charged plates

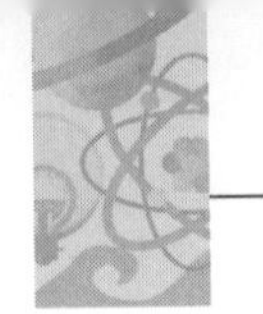

Around a point charge the field is 'radial', with field lines emanating like the spokes of a wheel. Between a pair of parallel conducting plates that have a potential difference applied across them, the field is 'uniform', with all field lines perpendicular to the plates. We can use these diagrams to assess the relative field strengths at different places in the field. For example, near the point charge the field lines are more concentrated and so the field is strong; further from the point charge the lines spread out, indicating a weakening of the field strength. Between the plates the lines of force are parallel and equally spaced, and the field strength is the same at all positions in the field.

Electric field strength

$$E = \frac{F}{Q} \qquad \text{units: N C}^{-1}$$

where F is the force that would act on a positive charge Q placed in the electric field.

It should be noted that E is a *vector* quantity whose direction is that of the force F.

Worked example

An electron gun in a cathode-ray tube has an electric field of strength $2.5 \times 10^5 \text{N C}^{-1}$ between the cathode and the anode. Electrons from the heated cathode are accelerated across the 2.0 cm gap between the electrodes; some pass through the hole in the anode and travel at constant speed until they strike a fluorescent screen.

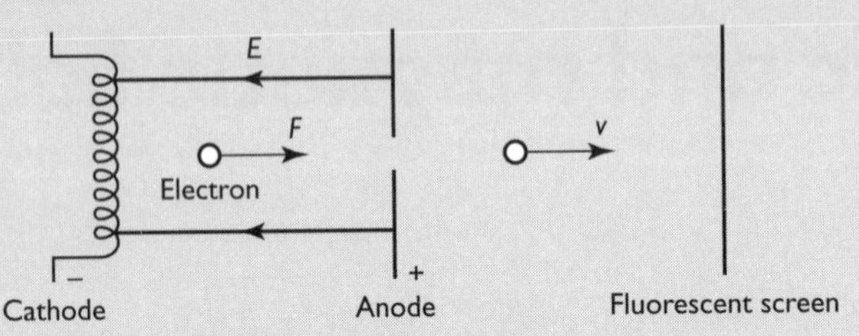

(a) Calculate the force on an electron in the field.
(b) Calculate the acceleration of the electron when it is between the electrodes.
(c) Assuming that the initial speed of the electron when it leaves the cathode is zero, calculate its speed when it leaves the gun.

Answer

(a) $F = EQ = 2.5 \times 10^5 \text{N C}^{-1} \times 1.6 \times 10^{-19} \text{C} = 4.0 \times 10^{-14} \text{N}$

(b) $a = \frac{F}{m} = \frac{4.0 \times 10^{-14} \text{N}}{9.1 \times 10^{-31} \text{kg}} = 4.4 \times 10^{16} \text{m s}^{-2}$

(c) $v^2 = u^2 + 2as = 0 + 2 \times (4.4 \times 10^{16} \text{m s}^{-2}) \times (2.0 \times 10^{-2} \text{m})$

so $v = 4.2 \times 10^7 \text{m s}^{-1}$

Tip Notice that the values of charge and mass of an electron are not provided in the question. In the examination, you can find these values in the list of physical constants at the end of the paper. This example also illustrates how you might be expected to use the equations of motion from Unit 1 in later units.

Coulomb's law

In the section on DC electricity in Unit 2, you will have performed simple experiments with statically charged rods to demonstrate the nature of forces between charges:

Like charges repel; unlike charges attract.

Coulomb's law shows how the magnitude of the electric force between two interacting charges depends on the magnitude of the charges and their distance apart:

$$F = \frac{kQ_1Q_2}{r^2}$$

where Q_1 and Q_2 are the magnitudes of the two point charges and r is the distance between them. The proportionality constant k is given by:

$$k = \frac{1}{4\pi\varepsilon_0}$$

where ε_0 is the **permittivity of free space** and has value $8.85 \times 10^{-12}\,\mathrm{F\,m^{-1}}$ (this is included in the data sheet at the end of each examination paper). In calculations the value $9.0 \times 10^9\,\mathrm{N\,m^2\,C^{-2}}$ is often used for k (the more precise value $8.99 \times 10^9\,\mathrm{N\,m^2\,C^{-2}}$ is given in the data sheet).

Worked example

Two identical charged spheres are suspended from a point, as shown in the diagram. Each sphere carries a charge of +50 nC, and the spheres are 10 cm apart. The threads are at an angle of 30°.

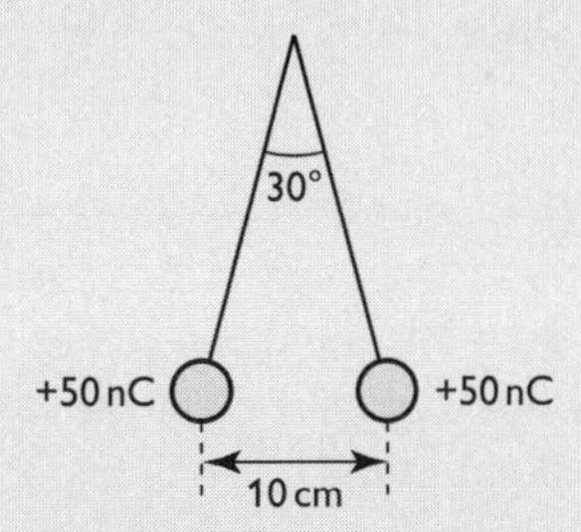

(a) Calculate the force between the two spheres.

(b) Draw a free-body force diagram for one of the spheres.

(c) Calculate the tension in each thread, and hence determine the mass of each sphere.

Answer

(a) $F = \dfrac{kQ_1Q_2}{r^2} = \dfrac{(9.0 \times 10^9\,\mathrm{N\,m^2\,C^{-2}}) \times (50 \times 10^{-9}\,\mathrm{C}) \times (50 \times 10^{-9}\,\mathrm{C})}{(10 \times 10^{-2}\,\mathrm{m})^2}$

$= 2.25 \times 10^{-3}\,\mathrm{N} \approx 2.3 \times 10^{-3}\,\mathrm{N}$

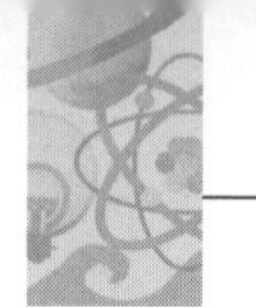

(b)

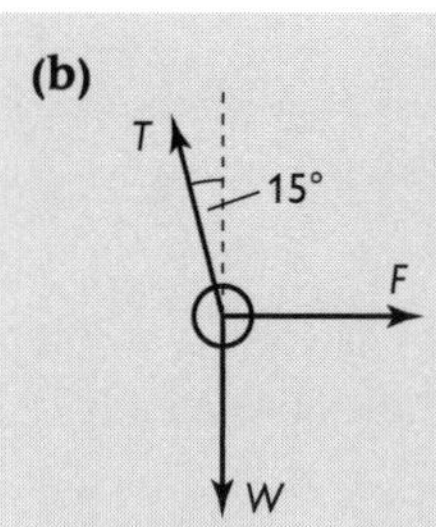

(c) Taking horizontal components:

$$T\sin 15° = 2.25 \times 10^{-3}\,\text{N so } T = \frac{2.25 \times 10^{-3}\,\text{N}}{\sin 15°} = 8.7 \times 10^{-3}\,\text{N}$$

Taking vertical components:

$$mg = T\cos 15° \text{ so } m = \frac{8.7 \times 10^{-3}\,\text{N} \times \cos 15°}{9.8\,\text{m s}^{-2}} = 8.6 \times 10^{-4}\,\text{kg} = 0.86\,\text{g}$$

The standard model of a hydrogen atom has an electron in a circular orbit around a proton. The centripetal force required to keep the electron in orbit is provided by the electrostatic force between the proton and the electron:

$$\frac{kQ_1Q_2}{r^2} = F = \frac{mv^2}{r}$$

The proton and the electron both have charge of magnitude 1.6×10^{-19} C; the mass of the electron is 9.1×10^{-31} kg and the radius of a hydrogen atom is 0.11 nm. Using these data, you should be able to show that the speed of the orbiting electron is about $1.5 \times 10^{6}\,\text{m s}^{-1}$.

Radial electric field strength

Recall the diagram of field lines for an electric field around a point charge. It was stated that the spreading out of the lines indicates that the field strength decreases with distance from the charge. An expression for the field strength around a point charge can be derived using Coulomb's law.

Consider the force F acting on a charge q at a distance r away from a point charge Q. We have:

$$F = \frac{kQq}{r^2} \quad \text{and} \quad E = \frac{F}{q}$$

It follows that:

$$E = \frac{kQ}{r^2}$$

so the field strength decreases with distance according to an inverse square law.

Uniform electric field strength

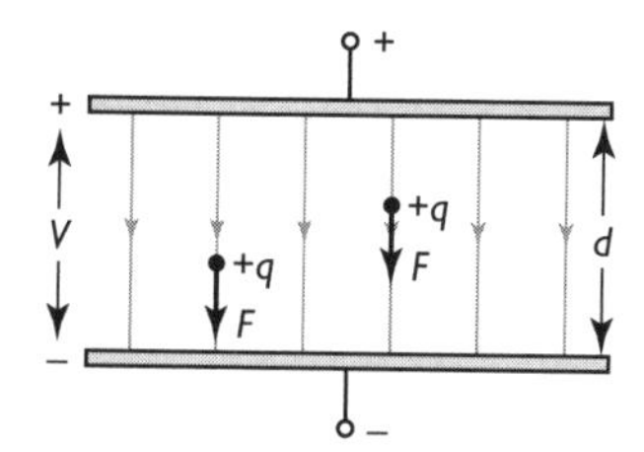

$F = Eq$ is the same at all points in the field

Between a pair of parallel plates with a potential difference applied across them, the electric field is uniform and so the force on a charge anywhere in the field is constant. Suppose that a small charge q is moved from one plate to the other; then the work done equals force × distance. On the other hand, from the definition of potential difference (in Unit 2), the work done in moving the charge is $V \times q$. Thus:

$$W = Fd = Vq$$

and hence $\frac{F}{q} = \frac{V}{d}$ or, in other words:

$$E = \frac{V}{d}$$

Worked example

(a) Show that the units NC^{-1} and Vm^{-1} are consistent.

(b) Calculate the electric field strength at a distance of 10 cm from a point charge of 800 nC.

(c)

+ 1.05 kV

0.5 cm

F

W

– 0 V

A charged oil drop is held in a fixed position between a pair of parallel plates. The plates are 0.50 cm apart and have a potential difference of 1.05 kV across them. The oil drop has a radius of 1.2×10^{-6} m and a density of $960\, kg\, m^{-3}$.

(i) Determine the field strength between the plates.

(ii) Calculate the charge on the oil drop.

Answer

(a) $V \equiv JC^{-1}$ and $J \equiv Nm$, so $Vm^{-1} \equiv (NmC^{-1})m^{-1} \equiv NC^{-1}$

(b) $E = \frac{kQ}{r^2} = \frac{(9.0 \times 10^9\, Nm^2 C^{-2}) \times (800 \times 10^{-9}\, C)}{(10 \times 10^{-2}\, m)^2} = 7.2 \times 10^5\, NC^{-1}$

(c) (i) $E = \frac{V}{d} = \frac{1.05 \times 10^3\,\text{V}}{0.50 \times 10^{-2}\,\text{m}} = 2.1 \times 10^5\,\text{N C}^{-1}$

(ii) For equilibrium:
electric force = weight of oil drop

$$Eq = mg$$

$$q = \frac{mg}{E} = \frac{\frac{4}{3}\pi(1.2 \times 10^{-6}\,\text{m})^3(960\,\text{kg m}^{-3}) \times 9.8\,\text{m s}^{-2}}{2.1 \times 10^5\,\text{N C}^{-1}}$$

$$= 3.2 \times 10^{-19}\,\text{C}$$

Capacitance

Capacitors

A capacitor is a device that can store charge. Any isolated conductor can behave as a capacitor, although most commercial capacitors consist of a pair of parallel plates separated by an insulating medium.

The ability of a capacitor to store charge depends on its dimensions, the nature of the insulating material and the potential difference applied across the plates. This is expressed as the **capacitance** C of the capacitor:

$C = \frac{Q}{V}$ unit: farad (F)

where Q is the charge stored when the capacitor is charged to a potential V.

Worked example

Calculate the charge stored on a 220 μF capacitor when the voltage applied is 12 V.

Answer

$Q = CV = 220 \times 10^{-6}\,\text{F} \times 12\,\text{V} = 2.6 \times 10^{-3}\,\text{C} = 2.6\,\text{mC}$

Note: Most capacitors are only able to store small charges. Capacitance values in the order of nF (10^{-9} F) or pF (10^{-12} F) are common in electronic circuits.

Measuring charge

Charge can be measured using a **coulometer**.

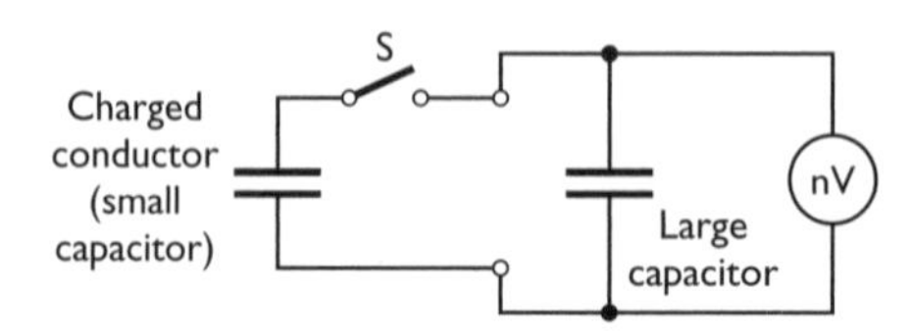

A coulometer consists of a capacitor whose known value of capacitance must be much larger than that of the conductor from which the charge is taken. For example, when a charged 1 F capacitor is connected to an uncharged 100 F capacitor, the charge will be shared between them, with 100 times more going to the 100 F capacitor — in effect, virtually all of the charge is transferred to the larger capacitor. By measuring the voltage across the known capacitor, the charge can be calculated from $Q = CV$. In practice, because the charge, and hence the voltage, is usually extremely small, a DC amplifier is used and the voltmeter is calibrated with the appropriate coulomb scale.

Energy stored in a capacitor

If a charged capacitor is discharged through a lamp or motor, the energy to light the lamp or drive the motor is stored on the capacitor.

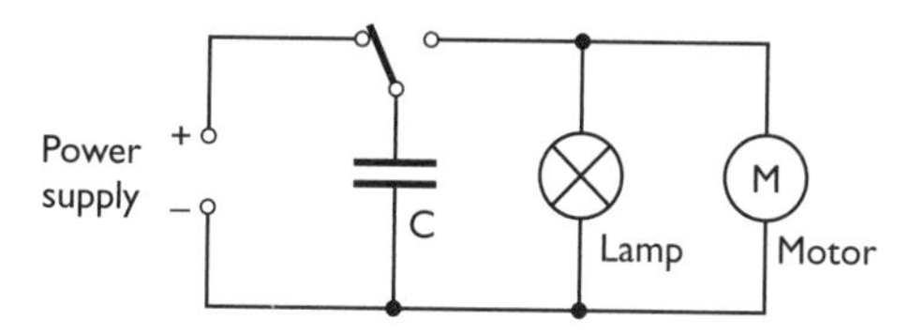

From the $Q = CV$ relation, a graph of V against Q for a capacitor will be a straight line through the origin. Suppose that a small charge δq is added to the capacitor when it is at voltage V^*; then the work done will be $V^*\delta q$, which is the area of the thin strip of width δq and height V^* shown in the following diagram. The total area beneath the line up to charge Q can be thought of as the sum of the areas of many such thin strips, and this sum of areas is equivalent to the total work done in charging the capacitor up to that point — that is, the energy stored in the capacitor.

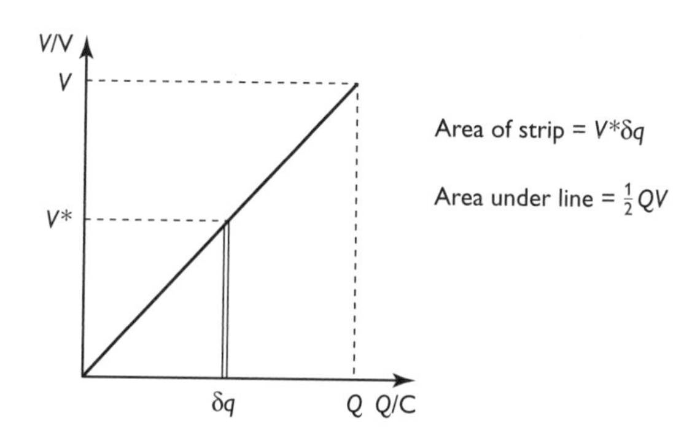

From the graph:

$$E = \text{area under the line up to } (Q, V) = \frac{1}{2}QV$$

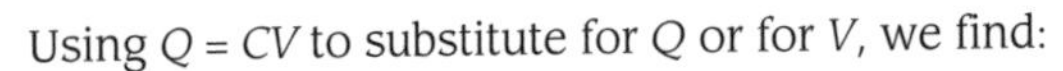

Using $Q = CV$ to substitute for Q or for V, we find:

$$E = \frac{1}{2}CV^2 = \frac{1}{2}Q^2/C$$

Worked example

A 2200 μF capacitor is charged to 12 V.

(a) Calculate the energy stored on the capacitor.

(b) When connected to a motor with a 20 g mass suspended from a thread wound around the spindle, the capacitor drives the motor, raising the mass through 30 cm. Determine the efficiency of the system.

Answer

(a) $E = \frac{1}{2}CV^2 = \frac{1}{2} \times (2200 \times 10^{-6}\,\text{F}) \times (12\,\text{V})^2 = 0.16\,\text{J}$

(b) Work done = $mgh = (20 \times 10^{-3}\,\text{kg}) \times 9.8\,\text{m s}^{-2} \times 0.30\,\text{m} = 0.059\,\text{J}$

Efficiency = $\frac{0.059\,\text{J}}{0.16\,\text{J}} \times 100\% = 37\%$

Charge and discharge of capacitors

When a capacitor is charged or discharged through a *resistor*, the charge (as well as the voltage) rises or falls *exponentially*.

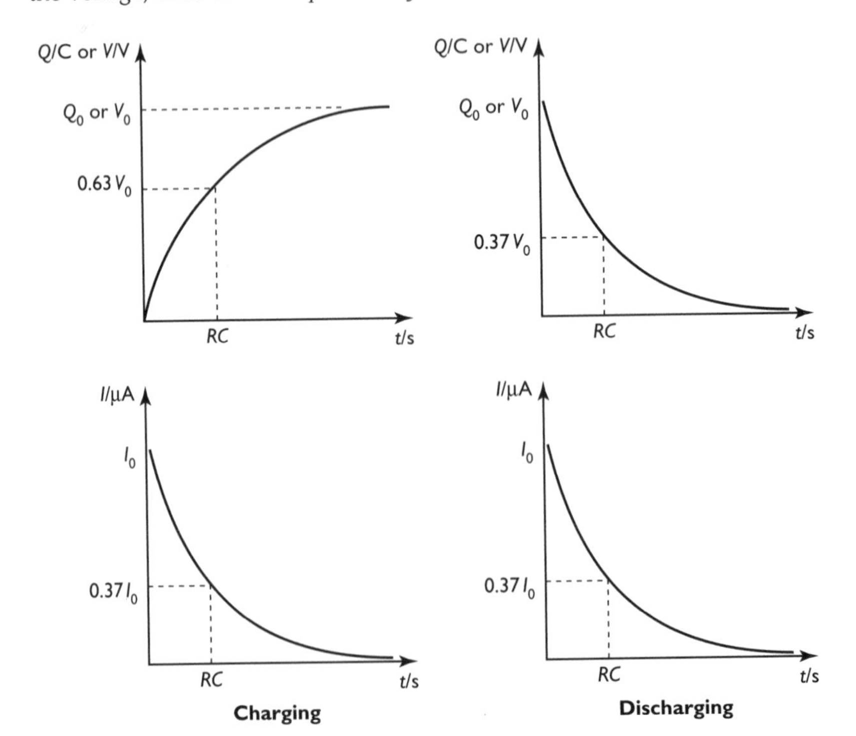

The charge (or voltage) against time graphs show that the growth or decay of charge is rapid initially, but then the rate reduces as the capacitor becomes charged or discharged.

For a *discharging* capacitor:

$$Q = Q_0 e^{-t/RC} \quad \text{and} \quad V = V_0 e^{-t/RC}$$

You will not be required to use the formula for charging a capacitor, $Q = Q_0(1 - e^{-t/RC})$.

For both charge and discharge of a capacitor, the gradient of the Q against t graph decreases in magnitude with time. This gradient represents the current flowing on or off the capacitor, and we have:

$$I = I_0 e^{-t/RC} \quad \text{(for both charging and discharging)}$$

From the exponential term $e^{-t/RC}$ we can see that the rate at which a capacitor charges or discharges through a resistor depends on the values of C and R. If the product RC is large, the decay will be slow; RC is known as the **time constant** of the circuit.

When $t = RC$:

$$Q = Q_0 e^{-RC/RC} = Q_0 e^{-1} \approx 0.37Q_0$$

Thus, over a time interval of duration equal to one time constant, the charge (also the voltage and the current) will fall to 37% of the initial value. This fact is the basis of most electronic timing circuits (e.g. the circuit controlling the time delay for a camera) — the voltage across a discharging capacitor falls until a certain switching threshold is reached.

Worked example

(a) Show that the unit of the time constant for a capacitor discharging through a resistor is the second.

A capacitor is charged to a voltage of 10.0 V and then discharged through a 100 kΩ resistor. Measurements of the voltage, taken every 10 s, are shown in the table below.

t/s	0	10	20	30	40	50	60
V/V	10.0	6.3	4.0	2.6	1.6	1.0	0.7

(b) Plot a graph of voltage against time for the discharging capacitor.

(c) Use your graph to determine the time constant for the circuit. Hence calculate the value of the capacitance of the capacitor.

Answer

(a) $RC = \frac{V}{I} \times \frac{Q}{V} = \frac{Q}{I}$, with units of $\frac{\text{C}}{\text{C s}^{-1}} = \text{s}$

(b)

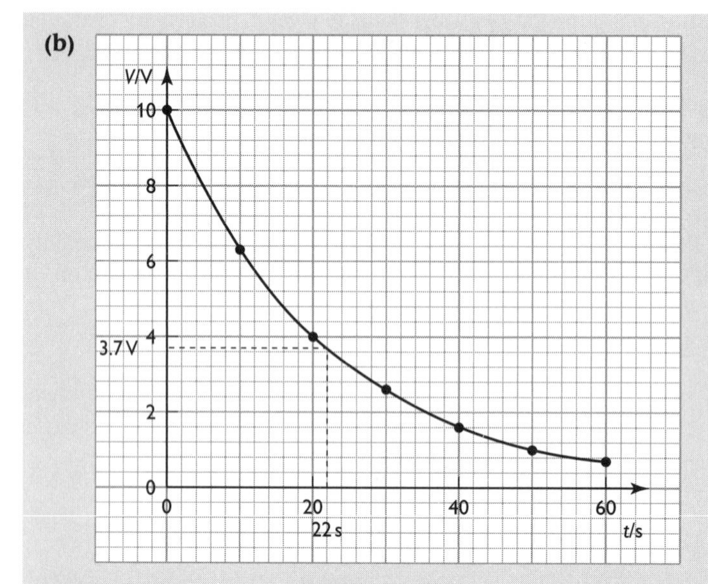

(c) The voltage will have fallen to 3.7 V (37% of the initial value 10.0 V) after one time constant, which, from the graph, is about 22 s.

So $RC = 22\text{ s} \Rightarrow C = \dfrac{22\text{ s}}{100 \times 10^3\ \Omega} = 2.2 \times 10^{-4}\text{ F} = 220\text{ F}$

Note: An alternative method of finding the time constant is to take the natural logarithm of the voltage and plot $\ln V$ against t. Using the laws of logarithms, $V = V_0 e^{-t/RC} \Rightarrow \ln V = \ln V_0 - \frac{t}{RC}$, so the graph of $\ln V$ against t will be a straight line whose gradient equals $\frac{-1}{RC}$.

Magnetic fields

A magnetic field is a region in which a magnetic force will be experienced. You are probably familiar with field patterns around bar magnets and coils carrying electric currents. However, unlike electric fields, it is less clear what is experiencing the force in this case.

A magnetic field is a region where a moving charge experiences a force.

The *moving charges* are usually in current-carrying wires, or in beams of electrons or ions.

Magnetic field strength (flux density)

The strength of a magnetic field is defined in terms of the force experienced by a current-carrying wire placed at right angles to the field.

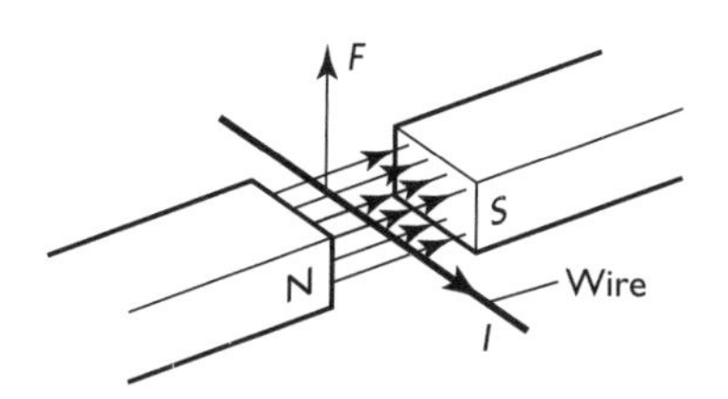

Fleming's left-hand rule gives the *direction* of the force acting on the wire with reference to the direction of the current and the direction of the field:

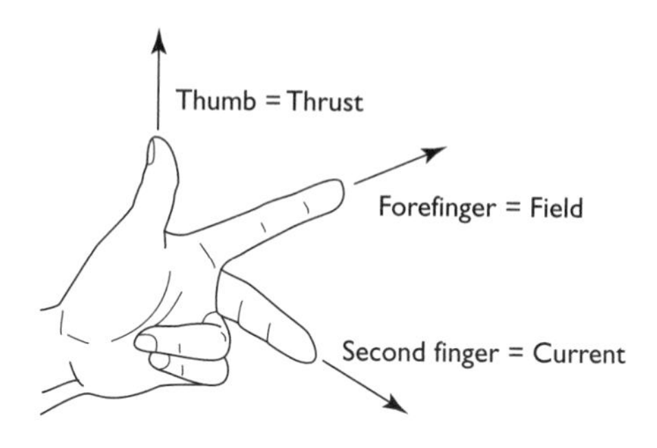

Experiments show that the *magnitude* of the force acting on the wire is proportional to the current I flowing in the wire and the length l of wire in the magnetic field, i.e. $F \propto Il$, or:

$$F = BIl$$

The constant B reflects the strength of the field and is called the **magnetic flux density** of the field.

$$B = \frac{F}{Il} \qquad \text{unit: tesla (T)}$$

If the wire is at an angle θ to the field, then the component of its length at right angles to the field is used to determine the magnitude of the force:

$$F = BIl\sin\theta$$

Worked example

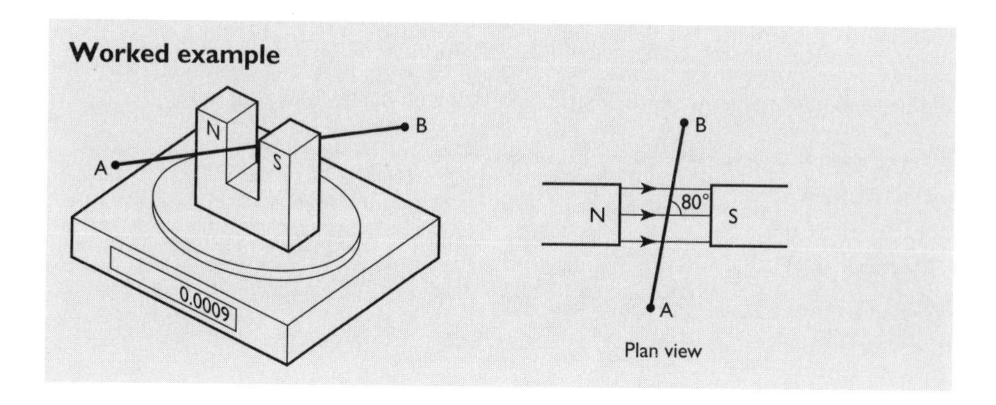

In an experiment to determine the force acting on a current-carrying wire in a magnetic field, a magnet is placed on a sensitive balance, and a stiff copper wire is rigidly clamped so that it passes between the poles. The balance is zeroed, and then a current of 2.0 A is passed through the wire. The balance now reads +1.110 g.

(a) What is the magnitude of the force exerted by the wire on the balance?

(b) State the direction in which the current is flowing.

(c) If the wire is at an angle of 80° to the field and has a length of 5.0 cm within the field, calculate the flux density between the poles of the magnet.

Answer

(a) $F = mg = (1.110 \times 10^{-3}\,\text{kg}) \times 9.81\,\text{m s}^{-2} = 1.1 \times 10^{-2}\,\text{N}$

(b) The wire exerts a downward force on the magnet, so, by Newton's third law, the magnet exerts an equal *upward* force on the wire. Using Fleming's left-hand rule, the current must flow from B to A.

(c) $B = \dfrac{F}{Il\sin\theta} = \dfrac{1.1 \times 10^{-2}\,\text{N}}{2.0\,\text{A} \times 0.050\,\text{m} \times \sin 80^\circ} = 0.11\,\text{T}$

Deflection of charged particles in magnetic fields

Moving charges constitute an electric current, and so will experience a force when they move at a non-zero angle to a magnetic field. Beams of electrons or ions will therefore be deflected as they pass through a magnetic field, as will fast-moving, charged subatomic particles.

The force on a particle carrying a charge q moving with velocity v at an angle θ to a field of flux density B is given by the expression:

$$F = Bqv\sin\theta$$

Note that since the force is always at right angles to the direction of motion of the charges, particles moving perpendicular to the field will travel along circular paths, with the magnetic force providing the centripetal acceleration. The identification of particles by observing the paths they take in a magnetic field is an important technique in particle physics, which we will look at in more detail later.

A comparison of magnetic and electric fields is given in the following table:

Electric field	Acts on both stationary and moving charges	The force acts in the direction of the field
Magnetic field	Acts only on moving charges	The force acts at right angles to the field

Worked example

A beam of electrons is fired between a pair of parallel plates that are 3.0 cm apart in a vacuum, as shown in the diagram.

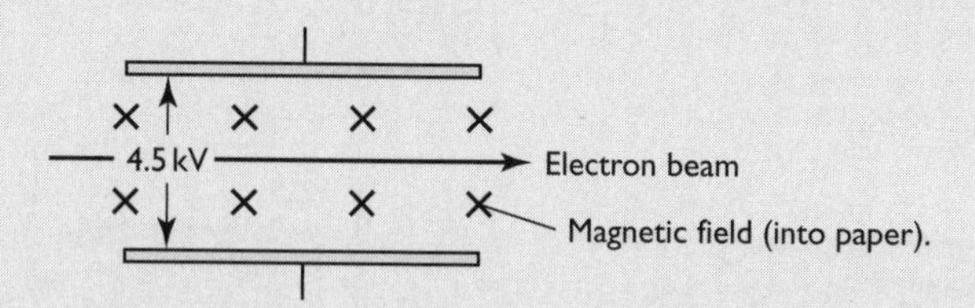

A potential difference of 4.5 kV is applied across the plates, and a uniform magnetic field of flux density 0.20 T is set up between the plates, at right angles to the direction of motion of the electrons.

(a) Determine the electric field strength between the plates, and hence calculate the force acting on an electron in this field.

(b) Write an expression for the magnetic force on an electron, and state the direction of the force if the magnetic field is as shown in the diagram (i.e. into the paper).

(c) Some electrons pass through the fields without being deviated. Explain why this is, and state the polarity of the voltage across the plates.

(d) Calculate the velocity of the undeviated electrons.

Answer

(a) $E = \frac{V}{d} = \frac{4.5 \times 10^3\,\text{V}}{3.0 \times 10^{-2}\,\text{m}} = 1.5 \times 10^5\,\text{V m}^{-1} = 1.5 \times 10^5\,\text{N C}^{-1}$

$F = Eq = 1.5 \times 10^5\,\text{N C}^{-1} \times 1.6 \times 10^{-19}\,\text{C} = 2.4 \times 10^{-14}\,\text{N}$

(b) $F = Bqv$ (the beam is at right angles to the magnetic field).
By Fleming's left-hand rule, the force is downward.

Note: For a flow of negatively charged electrons, the conventional current (as used in the rule) is in the *opposite* direction to the motion of the electrons, i.e. leftward in this case.

(c) Some electrons are not deviated because the electric and magnetic forces acting on them are equal and opposite. As the magnetic force is downward, the voltage across the plates must be such that the upper plate is positive, in order to generate an upward force on such electrons.

(d) Balancing the electric and magnetic forces:

$Eq = Bqv \Rightarrow v = \frac{E}{B} = \frac{1.5 \times 10^5\,\text{N C}^{-1}}{0.20\,\text{T}} = 7.5 \times 10^5\,\text{m s}^{-1}$

Flux and flux linkage

Magnetic fields are represented by 'field lines' or 'lines of force' in the same manner as electric fields. The lines are close together where the field is strong, and spread out as the field weakens; this can be seen in the following diagrams of magnetic fields around a bar magnet and a current-carrying solenoid.

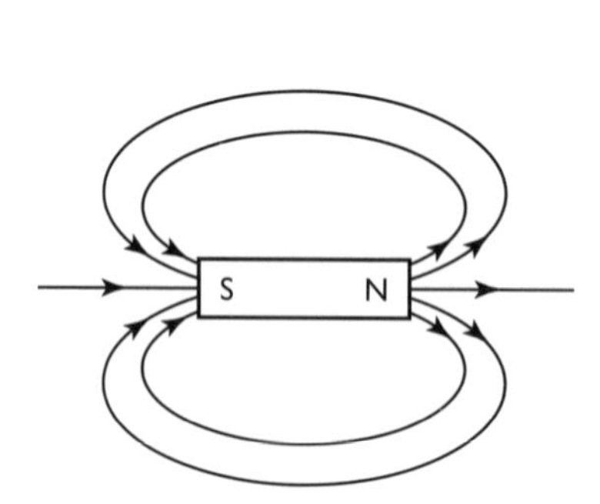

Field is strongest near the poles

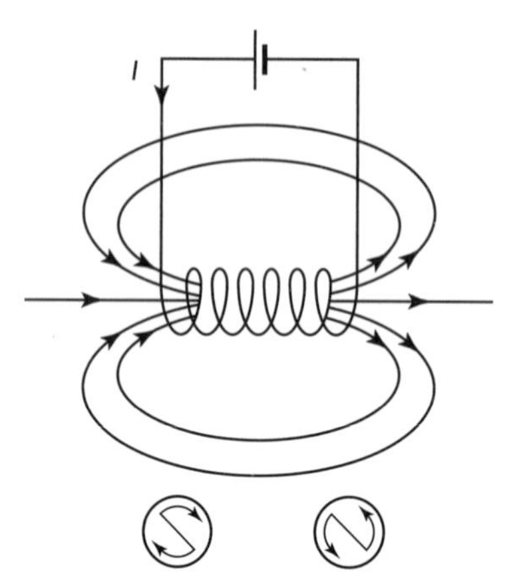

The end where the current flows in a clockwise sense behaves like the south pole of a magnet

The magnetic flux density B can be thought of as the concentration of field lines, or the concentration of the 'magnetic flux'. Earlier, the magnetic flux density was defined in terms of the force on a current-carrying wire, but it can also be thought of as the magnetic flux per unit area:

$$B = \frac{\Phi}{A}$$

where Φ denotes the flux and A is the area of a surface perpendicular to the magnetic field.

This leads to a definition of **magnetic flux** as:

$\Phi = BA$ unit: weber (Wb)

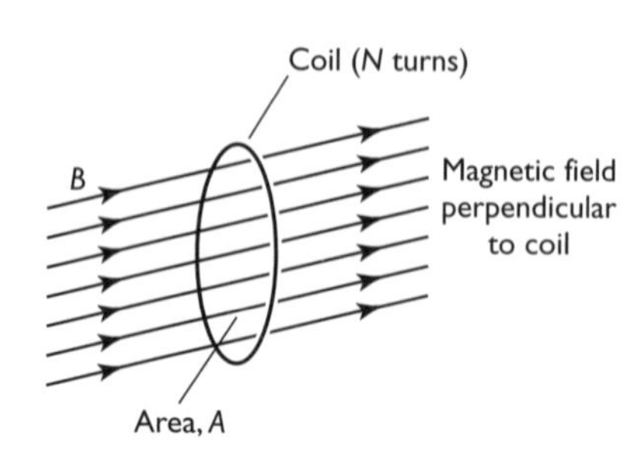

If a coil of N turns is placed with its edges perpendicular to a magnetic field, the total flux linked with the coil will be $N\Phi$. In other words:

flux linkage = $N\Phi = NBA$

Electromagnetic induction

Electromagnetic induction is the generation of a voltage in a conductor by the interaction of a changing magnetic field with the conductor. It can be simply demonstrated using magnets and coils of wire.

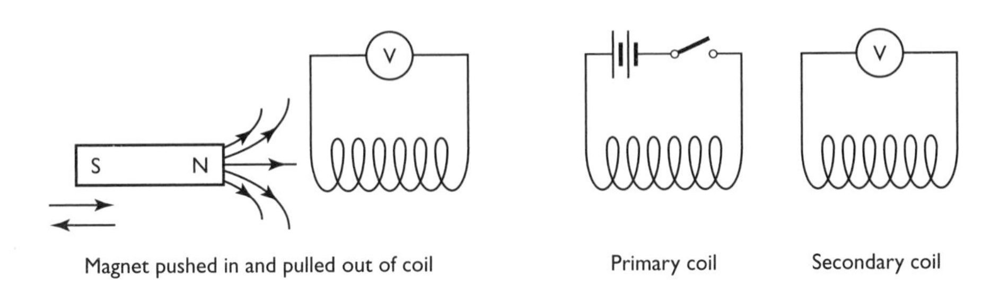

Magnet pushed in and pulled out of coil

Primary coil

Secondary coil

As the magnet is pushed into the coil, an emf is generated across the coil, which is displayed on the voltmeter. When the magnet is removed, an emf of the opposite polarity will be observed. It is important to realise that no emf will be induced while the magnet is stationary inside the coil — there needs to be a *change in flux linkage* for electromagnetic induction to take place. It is helpful to imagine the field lines of the magnet 'cutting through' the wire of the coil while the flux is changing.

Now suppose the magnet is replaced by a 'primary' coil connected to a power supply via a switch. When the circuit is switched on, a magnetic field will be formed and the flux will link with the secondary coil, thus generating an emf across the secondary coil. On switching off the circuit, the magnetic field collapses and the induced emf in the secondary coil will be of the opposite polarity. Once again, when the current in the primary coil is constant, there will be no change in flux linkage and therefore no emf induced across the secondary coil.

In the magnet–coil experiment, the *magnitude* of the induced *voltage* can be increased by:

- using a stronger magnet
- increasing the rate at which the magnet is inserted or removed from the coil
- increasing the number of turns in the coil

This can be understood in terms of **Faraday's law** of electromagnetic induction:

The magnitude of the induced emf is directly proportional to the rate of change of flux linked with the conductor.

The *direction* of flow of the induced *current* is a consequence of the law of conservation of energy. When the north pole of the magnet is pushed into the coil, work is done that transfers electrical energy to the circuit. The current must therefore flow in such a direction that it generates a magnetic field which opposes the entry of the magnet's north pole — the current in the coil should be flowing anticlockwise so that the end where the magnet is entering behaves like another 'north pole' to push against the magnet being inserted. When the magnet is withdrawn from the coil, the induced current creates a 'south pole' (with current flowing clockwise) at the end of the coil to oppose removal of the magnet.

This behaviour illustrates **Lenz's law** of electromagnetic induction:

The current induced in a conductor always flows in such a direction as to oppose the change producing it.

Faraday's and Lenz's laws combine to give a formula for the induced emf ε:

$$\varepsilon = -\frac{d(N\Phi)}{dt}$$

Tip Although this equation is usually written in differential form, all calculations in the examination will give you discrete values of flux linkage and time to work with. So it is worth remembering the formula as a word equation:

$$\text{emf} = \frac{\text{change in flux linkage}}{\text{time}} = -\frac{\Delta(N\Phi)}{\Delta t}$$

Worked example

A circular coil of radius 20 cm with 100 turns of wire is placed vertically with its faces along the north–south direction.

(a) If the horizontal component of the Earth's magnetic field has flux density 2.0×10^{-5} T, calculate the flux linked to the coil.

(b) The coil is rapidly rotated through 180° about its vertical axis. Determine the change in flux linkage.

(c) If the time taken for the rotation is 0.06 s, calculate the emf induced in the coil.

Answer

(a) Flux linkage = $N\Phi = NBA$

$$= 100 \times (2.0 \times 10^{-5}\,\text{T}) \times \pi(20 \times 10^{-2}\,\text{m})^2$$

$$= 2.5 \times 10^{-4}\,\text{Wb}$$

(b) After a 180° rotation the faces of the coil will be pointing in opposite directions relative to their original orientation, so the flux will be linked in the reverse direction: $N\Phi = -2.5 \times 10^{-4}$ Wb

Hence the change in flux linkage = $-2.5 \times 10^{-4}\,\text{Wb} - 2.5 \times 10^{-4}\,\text{Wb}$

$$= -5.0 \times 10^{-4}\,\text{Wb}$$

(c) Induced emf $= -\frac{\text{change in flux linkage}}{\text{time}} = -\frac{-5.0 \times 10^{-4}\,\text{Wb}}{0.06\,\text{s}} = 8.3 \times 10^{-3}\,\text{V}$

Applications of electromagnetic induction

There are many applications of electromagnetic induction in everyday life. Large rotating coils in magnetic fields are used to generate electricity. Alternating currents in the primary coils of transformers provide a continually changing flux that, when linked with different-sized secondary coils, can step up or step down the voltage. This is needed in the power transmission process, and for charging laptop computers and mobile phones. In motor car engines, the spark to ignite the fuel comes from rapidly collapsing the field linked to the ignition coil. This generates a high voltage that produces the spark in the spark plugs.

Lenz's law plays a role in electromagnetic braking, which is used for trains and heavy road vehicles. Electric cars use their motors as generators when slowing down. In addition to the braking effect, this has the advantage of using the kinetic energy of the car to charge up the battery.

Particle physics

This topic covers atomic structure, particle accelerators and the quark–lepton model of fundamental particles.

The nuclear atom

You will have learnt about the Rutherford model of an atom, which has the bulk of the matter contained in a relatively small nucleus consisting of protons and neutrons, with electrons orbiting around it.

Each element is identified by its **proton number** and each isotope by the total number of protons and neutrons — the **nucleon number**. Consider this equation:

$$^{238}_{92}\text{U} \rightarrow {}^{234}_{90}\text{Th} + {}^{4}_{2}\text{He}$$

Uranium has proton number 92 and nucleon number 238, so it has 92 protons and 146 neutrons. Upon emission of an alpha particle, which carries two protons and two neutrons, the proton number falls by two and the nucleon number by four, leaving thorium-234.

In all nuclear transformations, the total proton and nucleon numbers on both sides of the equation must balance.

Evidence for the nuclear atom

The evidence for Rutherford's nuclear model of an atom was provided by the large-angle alpha particle scattering experiment performed by Geiger and Marsden. Alpha particles were fired through a thin sheet of gold. Most of the particles passed through

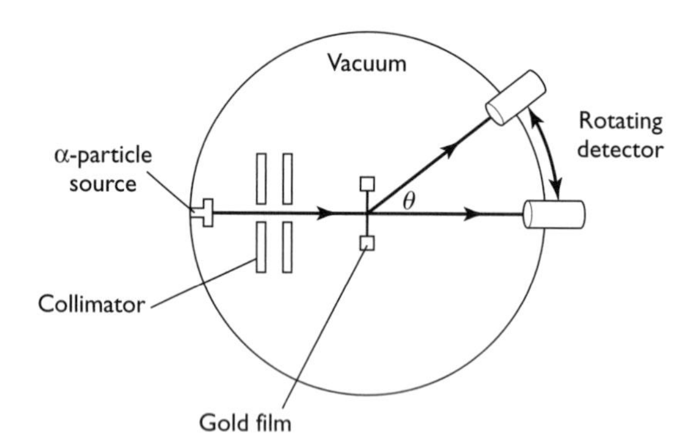

the film without being deviated, but a small number became scattered, including some that were deviated by large angles (greater than 90°).

In the examination, you may be asked to give full details of the experimental set-up; this information can be found in most textbooks. Below is an outline of the principles and major conclusions.

- Alpha particles are about 7500 times as massive as electrons, so their paths will be little affected by collisions with electrons.
- The diameter of a nucleus is very much less than the diameter of the atom; so, in a thin sheet of gold, the probability of electrons passing close to the nuclei is quite low.
- Alpha particles are positively charged ($+2e$, where e is the magnitude of charge on an electron), and so are the gold nuclei ($+79e$); therefore, the force between an alpha particle and a gold nucleus is repelling and will obey an inverse square law (Coulomb's law), decreasing rapidly with distance between the charged particles.
- If an alpha particle makes a direct collision with a nucleus, it can be scattered back towards the source.
- Detailed analysis of the number of particles scattered at a range of angles has supported the model of a small positive nucleus surrounded by a much larger envelope of orbiting electrons.

Worked example

An alpha particle of mass 6.7×10^{-27} kg moving at 1.0×10^{7} m s^{-1} is fired directly at the nucleus of a gold atom.

(a) Calculate the average force needed to bring the particle to rest, if it is stopped in a distance equal to the radius of the gold atom (1.3×10^{-10} m).

(b) Use Coulomb's law to determine the mean distance from the gold nucleus for such a force to act on the alpha particle.

(c) Comment on how your answer supports the experimental evidence for the nuclear atom.

Answer

(a) Using $v^2 = u^2 + 2as$, we have

$$a = \frac{v^2 - u^2}{2s} = \frac{(0\,\text{m s}^{-1})^2 - (1.0 \times 10^7\,\text{m s}^{-1})^2}{2(1.3 \times 10^{-10}\,\text{m})} = -3.8 \times 10^{23}\,\text{m s}^{-2}$$

Hence $F = ma = (6.7 \times 10^{-27}\,\text{kg}) \times (3.8 \times 10^{23}\,\text{m s}^{-2}) = 2.6 \times 10^{-3}\,\text{N}$

(b) From Coulomb's law,

$$r^2 = \frac{kQ_1Q_2}{F} = \frac{(9.0 \times 10^9\,\text{N m}^2\,\text{C}^{-2}) \times (2 \times 1.6 \times 10^{-19}\,\text{C}) \times (79 \times 1.6 \times 10^{-19}\,\text{C})}{2.6 \times 10^{-3}\,\text{N}}$$

and this gives $r = 3.7 \times 10^{-12}\,\text{m}$

(c) Because the force that a gold nucleus exerts on an alpha particle drops off sharply with distance according to an inverse square law, for the alpha particle to be stopped, it needs to pass extremely close to the gold nucleus. The distance found in part (b) is very small (about 3%) compared with the radius of the gold atom; this means that the nucleus must occupy a tiny region relative to the size of the atom, supporting the model of a nuclear atom.

Particle accelerators

Charged particles can be accelerated by electric and magnetic fields. One of the simplest particle accelerators is the **electron gun**, which is used in cathode-ray tubes and X-ray tubes. Electrons are produced from a heated metal filament or a metallic oxide surface by thermionic emission; then they are accelerated by an electric field.

Electron gun

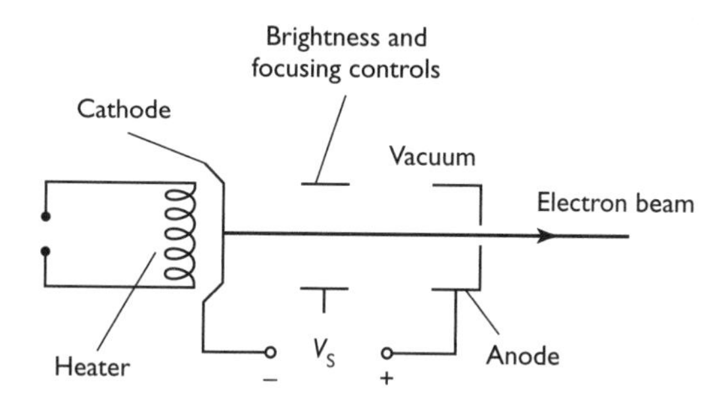

- A hot cathode emits electrons by thermionic emission (i.e. electrons near the surface gain sufficient thermal energy to overcome the work function; see Unit 2).
- The electrons experience a force in the electric field between the cathode and the anode, and are accelerated towards the anode:

  $F = Eq = ma$ (where q is the electron charge, $1.6 \times 10^{-19}\,\text{C}$)

- The work done on the electron by the field equals the gain in kinetic energy:

  $Vq = \frac{1}{2}mv^2$

- Some of the fast-moving electrons pass through the hole in the anode and strike a screen placed beyond.
- In practice, the guns may have several anodes that can have their voltages adjusted so as to focus the electron beam to a particular spot on the screen.

Worked example

(a) Show that the speed of an electron leaving an electron gun that has a potential difference of 10 kV between the cathode and the accelerating anode is about $6 \times 10^7\,\text{m s}^{-1}$.

(b) Some electrons may leave the gun with speeds greater than the above. Suggest how this is possible.

Answer

(a) From $Vq = \frac{1}{2}mv^2$ we have:

$(10 \times 10^3\,\text{V}) \times (1.6 \times 10^{-19}\,\text{C}) = \frac{1}{2}(9.1 \times 10^{-31}\,\text{kg})v^2$

So $v = 5.9 \times 10^7\,\text{m s}^{-1}$

(b) Some electrons will be emitted with extra kinetic energy gained from the process of thermionic emission.

Linear accelerator

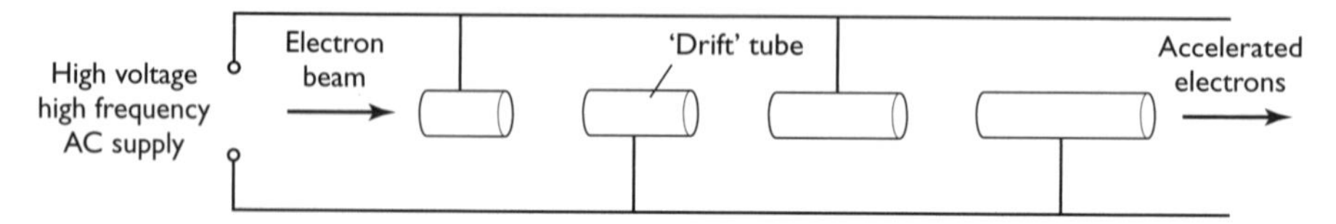

A linear accelerator (linac) operates on the same principle as the electron gun — electrons or other charged particles are accelerated across gaps between charged electrodes. In a linac, there are often as many as 100 000 'drift tubes' connected to a high-frequency, high-voltage AC supply. These are arranged in such a way that the particles gain kinetic energy between the tubes and move at constant speed inside the tubes. The main principles are:

- Alternate tubes are connected to each terminal of the AC supply.
- The charged particles spend one half of each period of the alternating voltage between two tubes and the other half of each cycle inside one of the tubes.
- During a half-cycle when the voltage would oppose their motion, the particles are inside a tube, where they are shielded from the electric field; the particles therefore travel at constant speed within the tube (i.e. they 'drift' through the tube).
- The particles gain kinetic energy as they travel across successive gaps, and can be accelerated to energies as high as 30 GeV.
- The length of the drift tubes increases along the accelerator so that although the speed of the particles is increasing, the time needed to pass through each tube will always be the same (equal to the half the period of the alternating electric field).

Cyclotron

The main disadvantage of linear accelerators is that, in order to produce high energies, they need to be very long. The Stanford linear accelerator in the USA has a length of about 3 km. Cyclotrons, while based on the same principle of synchronous acceleration as linacs, use a *magnetic* field to make the charged particles move in a spiral path.

To understand the action of a cyclotron, it might be helpful to look back at the section on deflection of charged particles in a magnetic field.

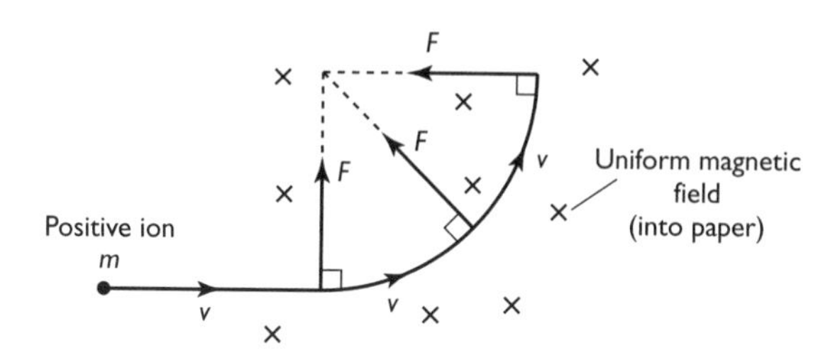

By Fleming's left-hand rule, a particle carrying charge q moving with speed v at right angles to a magnetic field of flux density B will experience a force of magnitude Bqv in a direction perpendicular to its motion. The particle will therefore follow a circular path, and we have $Bqv = F = \frac{mv^2}{r}$, so:

$$r = \frac{mv}{Bq} = \frac{p}{Bq}$$

This relationship $r = p/Bq$ between the momentum of a charged particle in a magnetic field and the radius of the circular motion also plays an important role in the identification of particles in detectors (see p. 43).

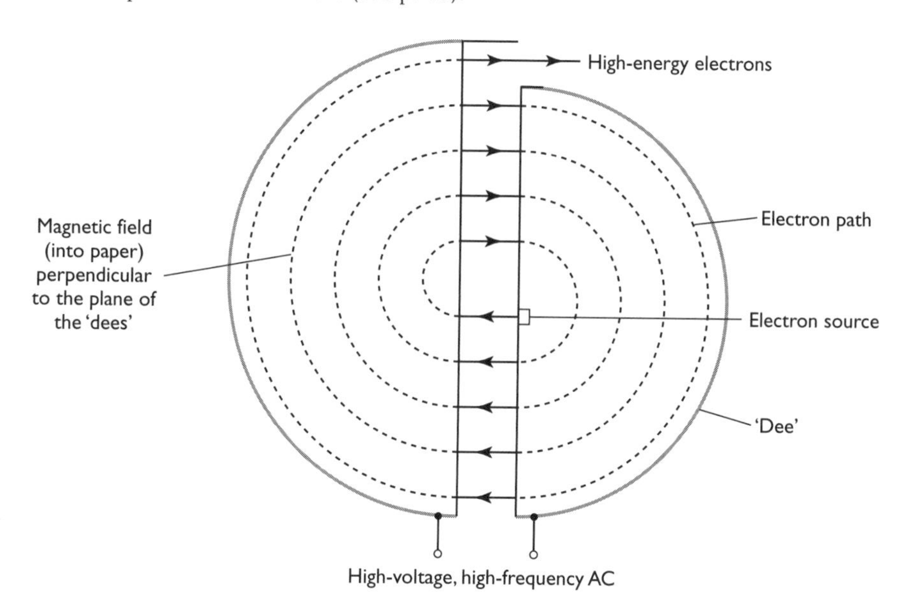

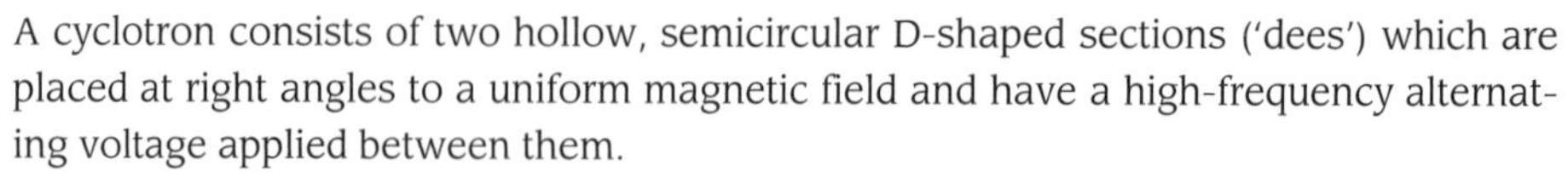

A cyclotron consists of two hollow, semicircular D-shaped sections ('dees') which are placed at right angles to a uniform magnetic field and have a high-frequency alternating voltage applied between them.

- An ion source fires charged particles into the gap between the dees close to the centre.
- The particles are accelerated across the gap during a half-cycle of the alternating voltage when the polarity is appropriate.
- During the next half-cycle, the particles follow a circular path at constant speed in one of the dees.
- The particles are then accelerated across the gap into the other dee.
- Faster particles follow paths of greater radius, so that all particles always spend the same amount of time in the dees and their acceleration is synchronised with the appropriate half-cycle of alternating voltage.

Worked example

A cyclotron has a maximum radius of 0.50 m and uses a magnetic field of strength 1.2 T. It is used to accelerate protons.

(a) Show that the momentum of the protons as they leave the cyclotron is about $1 \times 10^{-19}\,\mathrm{kg\,m\,s^{-1}}$.

(b) Calculate the kinetic energy of the emitted protons. Give your answer in MeV.

Answer

(a) $p = Bqr = 1.2\,\mathrm{T} \times (1.6 \times 10^{-19}\,\mathrm{C}) \times 0.50\,\mathrm{m} = 0.96 \times 10^{-19}\,\mathrm{kg\,m\,s^{-1}} \approx 1 \times 10^{-19}\,\mathrm{kg\,m\,s^{-1}}$

(b) $E_k = \dfrac{p^2}{2m} = \dfrac{(0.96 \times 10^{-19}\,\mathrm{kg\,m\,s^{-1}})^2}{2 \times (1.67 \times 10^{-27}\,\mathrm{kg})} = 2.8 \times 10^{-12}\,\mathrm{J}$

$$= \frac{2.8 \times 10^{-12}\,\mathrm{J}}{1.6 \times 10^{-19}\,\mathrm{J\,eV^{-1}}} = 1.7 \times 10^{7}\,\mathrm{eV} = 17\,\mathrm{MeV}$$

Relativistic effects

For speeds approaching the speed of light, relativistic effects need to be taken into account. For example, suppose that an electron is accelerated across a potential difference of 1 GV; if the electron has a mass of 9.11×10^{-31} kg (the value of electron mass listed on the data sheet), the equation $Vq = \frac{1}{2}mv^2$ predicts that v would be about $2 \times 10^{10}\,\mathrm{m\,s^{-1}}$. However, it is a basic postulate of the theory of relativity that nothing can travel beyond the speed of light ($3 \times 10^{8}\,\mathrm{m\,s^{-1}}$). Therefore, for the energy to be conserved, the *mass* of the electron must *increase*. The equivalence of mass and energy is discussed later; at this point you should just be aware that relativistic effects can create synchronisation problems in high-energy particle accelerators. **Synchrotrons**, such as CERN in Switzerland, account for these relativistic effects and can produce particles with extremely high energy.

Particle detectors

Particles can be detected when they interact with matter to cause ionisation or when they excite electrons to higher energy levels, accompanied by the emission of photons. Although specific details of particle detectors are not required for the examination, you are expected to know the basic principles and applications of ionisation chambers and semiconducting devices.

In bubble tanks and cloud chambers, a charged particle passing through will generate a trail of ions along which bubbles or vapour droplets are formed, making these paths visible. The nature of the particles can be deduced from the length of the trails they leave, and from how these paths are affected by electric and magnetic fields.

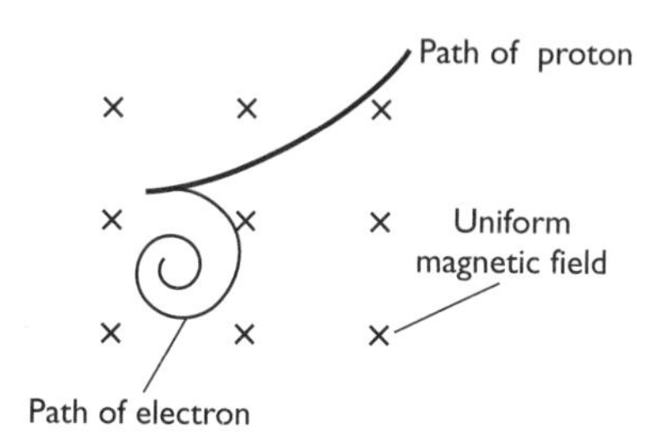

The diagram shows a bubble-tank image of the trails produced when a high-energy photon travelling from left to right strikes a neutron, which splits into a proton and an electron. The following information can be gained from the tracks:

- The photon is uncharged, so it does not produce a trail of ions and cannot be observed.
- The proton is positively charged, and Fleming's left-hand rule tells us that the force acting on the proton is initially upward. As the charge on the electron is negative, it will be deviated in the opposite direction.
- From the relation $r = \frac{p}{Bq}$ we can infer that the proton has greater momentum than the electron, because its path has a larger radius. Note that a thicker trail also indicates more intense ionisation, which is usually due to the particle having greater momentum or charge.
- The path of the electron spirals inward. This is because, as the electron loses energy, its momentum decreases and the radius of its path becomes smaller. Although less noticeable from the traces in the diagram, the radius of the proton's track will also decrease as its momentum reduces.

Examples of excitation detectors are fluorescent screens, scintillation counters and solid-state (electronic) devices.

Particle interactions

In all interactions, **energy**, **charge** and **momentum** must be conserved. In the bubble-tank example above, the particle interaction can be expressed as a simple equation:

$$n^0 + \gamma \rightarrow p^+ + e^-$$

It is clear that as both the neutron and the photon are uncharged, the initial charge is zero. After the interaction the total charge is also zero, because the proton carries a charge of $+e$ (where $e = 1.6 \times 10^{-19}$C is the magnitude of electron charge) and the electron a charge of $-e$.

The photon has momentum, so if the neutron was stationary before the interaction, the total momentum of the proton and the electron must be the same as the initial momentum of the photon.

The energy conservation in this scenario is less clear. The photon has energy hf, and the proton and electron will gain kinetic energy after the collision. However, the total mass of the proton and the electron differs from the mass of the neutron. Here, mass–energy equivalence must be accounted for.

The **rest mass** of the particles can be represented as an equivalent **energy** by means of Einstein's famous equation:

$$E_0 = m_0c^2 \qquad \text{or} \qquad m_0 = \frac{E_0}{c^2}$$

The rest mass can be thought of as the energy that would be transferred if the entire mass were to be dematerialised, or if the particle were to be made up completely from other forms of energy.

It is often convenient to represent the rest mass of subatomic particles in terms of the non-SI units MeV/c^2 or GeV/c^2.

- One electronvolt (eV) is the work done in moving an electron through a potential difference of one volt:
  $1\,\text{eV} = (1.6 \times 10^{-19}\text{C}) \times 1\,\text{V} = 1.6 \times 10^{-19}\text{J}$
- $1\,\text{MeV} = 1.6 \times 10^{-13}\text{J}$
  $1\,\text{GeV} = 1.6 \times 10^{-10}\text{J}$

MeV and GeV are units of energy; MeV/c^2 and GeV/c^2 are the corresponding units of mass.

Worked example

Express the rest mass of **(a)** an electron, and **(b)** a proton in terms of MeV/c^2.

Answer

(a) The rest mass energy of an electron is given by

$E_0 = m_0c^2 = 9.11 \times 10^{-31}\,\text{kg} \times (3.00 \times 10^8\,\text{m s}^{-1})^2 = 8.20 \times 10^{-14}\,\text{J}$

$= \dfrac{8.20 \times 10^{-14}\,\text{J}}{1.60 \times 10^{-19}\,\text{J eV}^{-1}} = 5.12 \times 10^5\,\text{eV} = 0.512\,\text{MeV}$

So the rest mass of an electron is $m_0 = 0.512\ \text{MeV}/c^2$

(b) The rest mass energy of a proton is

$E_0 = m_0c^2 = 1.67 \times 10^{-27}\,\text{kg} \times (3.00 \times 10^8\,\text{m s}^{-1})^2 = 1.50 \times 10^{-10}\,\text{J}$

$= \dfrac{1.50 \times 10^{-10}\,\text{J}}{1.60 \times 10^{-19}\,\text{J eV}^{-1}} = 9.39 \times 10^8\,\text{eV} = 939\,\text{MeV}$

So the rest mass of a proton is $m_0 = 939\ \text{MeV}/c^2$

Another unit used in particle physics is the **unified atomic mass unit**, denoted by 'u'. It is defined as one-twelfth of the mass of a carbon-12 atom, so:

$1\,\text{u} = 1.66 \times 10^{-27}\,\text{kg}$

The energy equivalent (m_0c^2) is 1.49×10^{-10} J or 931 MeV.

You will often see the concept of mass–energy equivalence expressed as:

$\Delta E = c^2 \Delta m$

where 'Δ' means 'the change in'. This is the form given in the formulae sheet.

Creation and annihilation of matter and antimatter

Early work with particle detectors showed that cosmic rays could produce some tracks identical to those of an electron, but which curve in the opposite direction. This was the first piece of evidence for the existence of **antimatter**. The underlying process of **pair production** is:

photon + photon → particle + antiparticle

An example of an antiparticle is the antielectron, or **positron**, whose mass is identical to that of the electron but which carries the opposite charge (of equal magnitude). Pair production in this case can be written as:

$\gamma + \gamma \rightarrow e^- + e^+$

As usual, it is necessary for momentum to be conserved in the creation of the electron–positron pair. Production of the positron ensures that the net charge remains zero. It is also essential that the gamma-ray photons have at least as much energy as the combined rest mass energy of the electron and the positron.

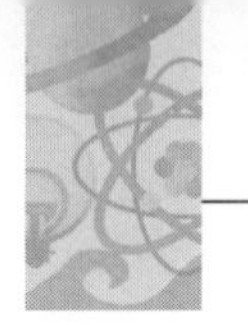

Worked example

Two gamma-ray photons interact to form an electron–positron pair.

(a) Calculate the minimum energy of the gamma rays that is required for this interaction.

(b) If the gamma rays both have a wavelength of 2.10×10^{-12} m, calculate the maximum kinetic energy of the electron and positron.

Answer

(a) A positron and an electron have the same rest mass energy, which we found to be 0.512 MeV in the previous worked example. So the minimum energy that the gamma rays need to have is:

$2 \times 0.512\,\text{MeV} = 1.02\,\text{MeV}$

(b) Energy of each photon $= \dfrac{hc}{\lambda} = \dfrac{(6.63 \times 10^{-34}\,\text{J s}) \times (3.00 \times 10^{8}\,\text{m s}^{-1})}{2.10 \times 10^{-12}\,\text{m}}$

$= 9.47 \times 10^{-14}\,\text{J}$

Energy of both photons $= 2 \times 9.47 \times 10^{-14}\,\text{J} = 1.89 \times 10^{-13}\,\text{J} = 1.18\,\text{MeV}$

Maximum kinetic energy of electron and positron

$= 1.18\,\text{MeV} - 1.02\,\text{MeV} = 0.16\,\text{MeV}$ (or 0.08 MeV per particle)

When matter collides with antimatter, we have the reverse of the pair production equation:

particle + antiparticle → photon + photon

Antiparticles have a very short lifetime. This is because matter is much more prevalent than antimatter, so antiparticles exist for only a brief time before they meet a particle and are **annihilated** to generate gamma-ray photons.

Subatomic particles

A subatomic particle is any particle that is smaller than an atom. **Fundamental particles**, also known as **elementary particles**, are subatomic particles that are thought to be indecomposable and which form the building blocks for other subatomic particles. There are two kinds of fundamental particles that you need to know about — **leptons** and **quarks**.

The leptons consist of the electron e^-, muon μ^- and tau particle τ^- (each carrying charge $-e$), the electron neutrino ν_e, muon neutrino ν_μ and tau neutrino ν_τ (each of charge zero), and their antiparticles.

The quarks consist of the particles d (down), u (up), s (strange), c (charm), b (bottom) and t (top); their antiparticles are called 'antiquarks'. The particles d, s and b carry charge $-\frac{1}{3}e$, while u, c and t carry $+\frac{2}{3}e$.

Hadrons are subatomic particles which are composed of quarks. They are classified into baryons and mesons.

A **baryon** is made up of three quarks. Baryons other than protons and neutrons are very short-lived. Protons carry charge $+e$ and are composed of the quarks uud; neutrons have charge zero and quark composition udd.

A **meson** is made up of a quark and an antiquark. For example, a π^+ meson is made from u and $\bar{d}$ (anti-d), and a π^- meson from d and $\bar{u}$ (anti-u); a K^+ meson is made from u and $\bar{s}$, and a K^- meson from s and $\bar{u}$.

The Edexcel specification does not require you to remember the exact composition of the various hadrons, but you are expected to be familiar with the notation and be able to interpret equations involving standard particle symbols such as π^+ and e^-.

Wave–particle duality

Diffraction patterns can be observed when high-energy electrons from an electron gun are fired through a thin slice of carbon, indicating that the electron, as well as being a particle, behaves like a wave. It is also possible to measure the 'radiation pressure' of light from the Sun, which shows that photons have momentum, an attribute of particles. This wave–particle duality (which is studied in detail in Unit 2) applies to all particles, but it is significant only on a subatomic scale. The relationship between the momentum p of a particle and its wavelength λ is expressed by **de Broglie's wave equation**:

$$\lambda = \frac{h}{p}$$

where h is the Planck constant.

In the electron diffraction experiment it was also observed that when the voltage across the electron gun is increased, the higher-energy electrons produced have a shorter wavelength. This relationship can be seen from Planck's photon equation (covered in Unit 2):

$$E = hf \quad \text{or} \quad E = \frac{hc}{\lambda}$$

Worked example

Calculate the de Broglie wavelength of:

(a) a tennis ball of mass 57 g travelling at 50 m s^{-1}

(b) an electron with kinetic energy 100 keV

Answer

(a) $\lambda = \frac{h}{p} = \frac{6.63 \times 10^{-34}\,\text{J s}}{0.057\,\text{kg} \times 50\,\text{m s}^{-1}} = 2.3 \times 10^{-34}\,\text{m}$

(b) $E_k = 100 \times 10^3\,\text{eV} \times 1.6 \times 10^{-19}\,\text{J eV}^{-1} = 1.6 \times 0^{-14}\,\text{J}$

From $E_k = \frac{p^2}{2m}$,

$$p = \sqrt{2mE_k} = \sqrt{2 \times (9.1 \times 10^{-31}\,\text{kg}) \times (1.6 \times 10^{-14}\,\text{J})} = 1.7 \times 10^{-22}\,\text{kg m s}^{-1}$$

$$\lambda = \frac{h}{p} = \frac{6.63 \times 10^{-34}\,\text{J s}}{1.7 \times 10^{-22}\,\text{kg m s}^{-1}} = 3.9 \times 10^{-12}\,\text{m}$$

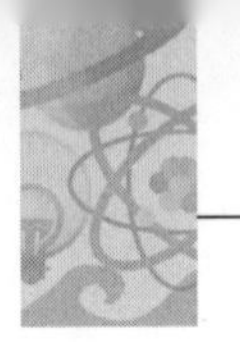

From the above example it can be seen that the wavelength of a 100keV electron is of the same order of magnitude as the diameter of a carbon atom, so a noticeable diffraction pattern is produced when an electron beam is focused on a specimen containing carbon. This principle is applied in electron microscopes.

If a high-energy accelerator is used, the wavelength of the electrons will be much shorter, and this enables the structure of hadrons to be probed using deep inelastic scattering.

Questions & Answers

The following two tests are made up of questions similar in style and content to those appearing in a typical Unit 4 examination. Each paper carries a total of 80 marks and should be completed in 1 hour 35 minutes. The first ten questions are multiple-choice objective tests with four alternative responses each. The remaining questions vary in both length and style, and the number of marks that each is worth ranges from 3 to 16. You might like to work through a complete paper in the allotted time and then check your answers; alternatively, you could separately attempt the multiple-choice section and selected longer questions to fit your revision plan. Use the fact that there are 95 minutes available for the 80 marks on the test to help you estimate how long you ought to spend on a particular question — you should be looking at about 10 minutes for the multiple-choice section, 5 minutes on a 4-mark short question, and approximately 15 minutes on a longer 12-mark question.

Although these sample papers resemble actual examination scripts in most respects, be aware that during the examination you will be writing your answers directly onto the paper, which is not possible for the tests in this book. It may be that you will need to copy diagrams and graphs that you would normally just write or draw onto in the real examination. If you are attempting one of these papers as a timed test, allow yourself an extra few minutes to account for this.

Remember that the values of physical constants such as the mass of an electron will not be provided in the test questions, but you can look them up in the data and formulae sheets. Many important physical relationships are also listed on these sheets, so it is well worth familiarising yourself with them.

The answers should not be treated as model solutions because they represent the bare minimum necessary to gain the marks. In some instances, the difference between an A-grade response and a C-grade response is suggested. This is not possible, however, for the multiple-choice section and many of the shorter questions that do not require extended writing.

Ticks (✓) are included in the answers to indicate where the examiner has awarded a mark. Note that half marks are not given.

Examiner's comments

Where appropriate, the answers are followed by examiner's comments, denoted by the icon *e*. These are interspersed in the answers and indicate where credit is due and where lower-grade candidates typically make common errors. They may also provide useful tips.

Test Paper 1

Questions 1–10

For questions 1–10, select one answer from A to D.

(1) The angular velocity of a particle moving with a tangential velocity of 2.0 m s^{-1} in a circular path of radius 20 cm is:

A 0.40 rad s^{-1}

B 0.80 rad s^{-1}

C 10 rad s^{-1}

D 20 rad s^{-1} (1 mark)

(2) Which of the following is the same unit as the tesla?

A A m N^{-1}

B N A m

C N A m^{-1}

D N A^{-1} m^{-1} (1 mark)

(3) An emf will be induced in a conducting wire if it is placed:

A parallel to a steady magnetic field

B parallel to a changing magnetic field

C perpendicular to a steady magnetic field

D perpendicular to a changing magnetic field (1 mark)

(4) A muon has a mass of about 106 MeV/c^2. Its mass in kilograms is approximately:

A 2×10^{-34}

B 2×10^{-28}

C 2×10^{-25}

D 2×10^{-22} (1 mark)

The following are four possible graphs of a quantity *Y* plotted against another quantity *X*. Refer to these graphs when answering questions 5, 6 and 7.

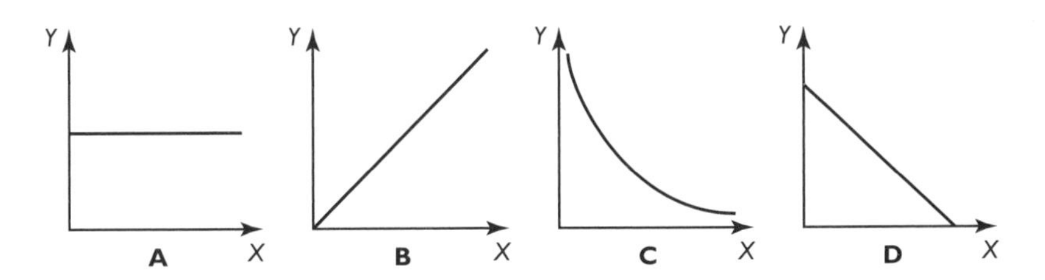

(5) Which graph *best* represents *Y* when it is the electric field strength close to a point charge and *X* is the distance from the charge? (1 mark)

(6) Which graph *best* represents *Y* when it is the force acting on a positively charged particle between two parallel plates with a constant potential difference across them and *X* is the distance from the positive plate? (1 mark)

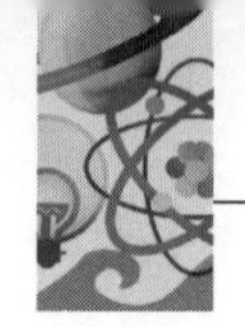

(7) Which graph *best* represents *Y* when it is the momentum of a proton moving at right angles to a uniform magnetic field and *X* is the radius of the proton's path? (1 mark)

(8) What is the energy stored on a 100 μF capacitor when the potential difference across it is 6.0 V?

A 0.30 mJ

B 0.60 mJ

C 1.80 mJ

D 3.60 mJ (1 mark)

(9) The momentum of an object of mass 0.12 kg moving with kinetic energy 24 J is:

A 2.4 kg m s^{-1}

B 4.8 kg m s^{-1}

C 12 kg m s^{-1}

D 24 kg m s^{-1} (1 mark)

(10) A K^- meson is composed of which combination of quarks?

A us

B $\bar{u}$s

C dss

D uss (1 mark)

Answers to Questions 1–10

(1) C

$\omega = \frac{v}{r} = \frac{2.0\,\text{m s}^{-1}}{0.20\,\text{m}} = 10\,\text{rad s}^{-1}$

(2) D

This comes from $B = \frac{F}{Il}$.

(3) D

It is important to remember that an emf cannot be induced in a conductor unless there is a *change* in flux linkage; this requires a changing field at right angles to the conductor.

(4) B

$m_0 = \frac{E_0}{c^2} = \frac{(106 \times 10^6\,\text{eV}) \times (1.6 \times 10^{-19}\,\text{J eV}^{-1})}{(3.0 \times 10^8\,\text{m s}^{-1})^2} = 1.9 \times 10^{-28}\,\text{kg}$

(5) C

The inverse square relationship $E = \frac{kQ}{r^2}$ is best represented by the curve in C.

(6) A

This set-up gives a uniform electric field. The field strength is constant, so the force ($F = Eq$) on the charged particle will be the same at all points between the plates.

(7) B

The radius of the proton's path is given by $r = \frac{p}{Bq}$. As the charge and the field strength are both constant, the momentum is directly proportional to the radius and so the graph is a straight line through the origin.

(8) C

$E = \frac{1}{2}CV^2 = \frac{1}{2}(100 \times 10^{-6}\,\text{F}) \times (6.0\,\text{V})^2 = 1.8 \times 10^{-3}\,\text{J} = 1.8\,\text{mJ}$

(9) A

$E_k = \frac{p^2}{2m} \Rightarrow p = \sqrt{2mE_k} = \sqrt{2 \times 0.12\,\text{kg} \times 24\,\text{J}} = 2.4\,\text{kg}\,\text{m}\,\text{s}^{-1}$

(10) B

A meson is made from a quark and an antiquark, so the only possible answer is B. You can check that the charges $\left(-\frac{2}{3}\text{e} \text{ for } \bar{\text{u}} \text{ and } -\frac{1}{3}\text{e for s}\right)$ add up to −1e, which is correct for the K^- meson. Note that the specification does not require you to remember the quark composition of a K^- meson — however, you are expected to know that particles made up of three quarks are baryons; so even though uss also has a charge of −1e, it cannot be the right answer.

Question 11

A fairground ride includes a loop in the form of a vertical circle of radius 5.0 m.

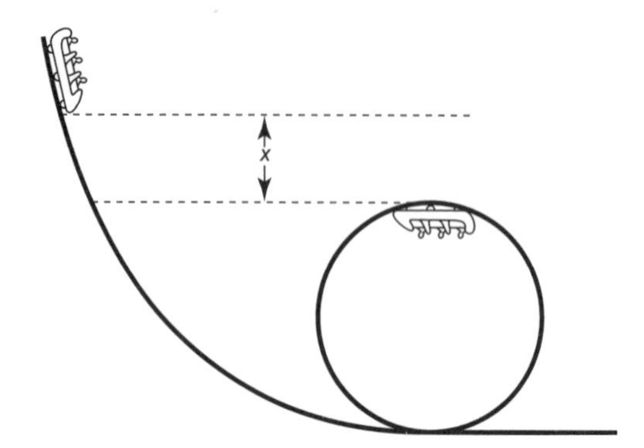

(a) Show that the minimum speed of the cart needed to ensure that the wheels stay in contact with the track at the top of the loop is 7 m s^{-1}. (2 marks)

(b) Calculate the minimum height, *x*, above the top of the loop through which the cart must fall in order that it completes the loop. (2 marks)

(c) In practice, the cart always falls from a height which is much bigger than that calculated in part (b). Explain why this is necessary. (1 mark)

Total: 5 marks

Answer to Question 11

(a) The centripetal force is provided by the weight: $\frac{mv^2}{r} = mg$ ✓

so $v = \sqrt{gr} = \sqrt{9.8\,\text{m s}^{-2} \times 5.0\,\text{m}} = 7.0\,\text{m s}^{-1}$ ✓

(b) Loss in gravitational potential energy = gain in kinetic energy *mgx*

$$mgx = \frac{1}{2}mv^2$$

$$x = \frac{v^2}{2g} = \frac{(7.0\,\text{m s}^{-1})^2}{2 \times 9.8\,\text{m s}^{-2}} \checkmark$$

$$= 2.5\,\text{m} \checkmark$$

(c) Some of the gravitational potential energy will be converted into other forms of energy because of friction, air resistance etc. ✓

Question 12

Beta-plus decay can be represented by the equation:

$p \rightarrow n + e^{+} + \nu_e$

(a) Identify the four particles in the reaction, and state which are baryons and which are leptons. (3 marks)

(b) Explain how the quark structure of the baryons changes in this process. (2 marks)

Total: 5 marks

Answer to Question 12

(a) p is a proton and n is a neutron; these are baryons. ✓
e^{+} is a positron and ν_e is an electron neutrino; ✓ both are leptons. ✓

e A grade-C candidate may lose a mark by just stating 'neutrino'. You are expected to be familiar with the fundamental particles and their symbols — including the three different kinds of neutrinos. Antiparticles should also be recognised (and the term 'antielectron' would be accepted in place of the much more commonly used 'positron').

(b) proton (uud) → neutron (udd) ✓
An up quark is converted to a down quark. ✓

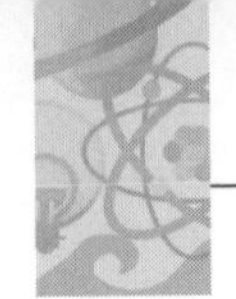

Question 13

Linear accelerators are used to produce high-energy particles. The diagram shows the structure of part of such a device.

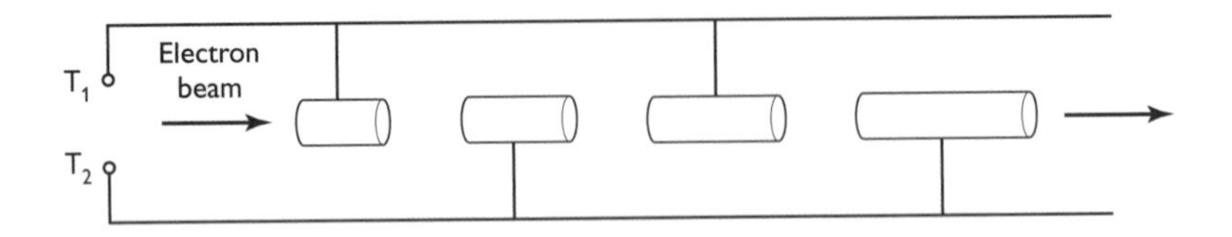

(a) State the nature of the power supply connected to terminals T_1 and T_2. (2 marks)

(b) Explain why the electrons travel with constant velocity when they are inside the tubes. (2 marks)

(c) Explain why the tubes increase in length along the accelerator. (2 marks)

(d) The Stanford linear accelerator is capable of accelerating electrons to energies of 30 GeV. Convert this energy into joules. (1 mark)

(e) State two reasons why such high-energy particles are needed for investigating fundamental particles. (2 marks)

Total: 9 marks

Answer to Question 13

(a) Alternating (AC) supply ✓ of high voltage and high frequency. ✓

(b) There is no electric field inside the tubes (every part of each tube is at the same voltage at any instant), ✓ so no force acts on the electrons while they are inside a tube. ✓

(c) The speed of the electrons increases along the accelerator. ✓ Therefore the tubes need to get longer so that the time the electrons spend in each stays the same. ✓

(d) $30\,\text{GeV} = 30 \times 10^{9}\,\text{eV} \times 1.6 \times 10^{-19}\,\text{J eV}^{-1} = 4.8 \times 10^{-9}\,\text{J}$ ✓

(e) *Any two of the following reasons*:

High energies are needed to overcome repulsive forces or to break up particles into their constituents. ✓

Sufficient energy is required for the rest mass energy of new particles created. ✓

Higher energies mean shorter wavelengths $\left(\text{from } E = \frac{hc}{\lambda}\right)$, and short wavelengths are needed to examine fine structure. ✓

Question 14

(a) State the principle of conservation of linear momentum. (2 marks)

A student investigating the law uses an air track with two gliders as shown in the diagram:

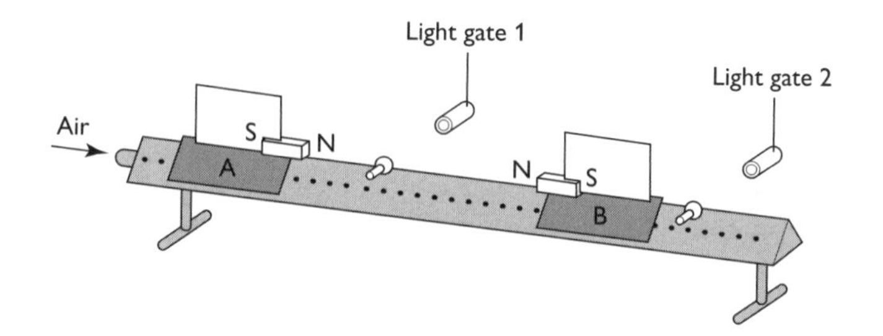

The student pushes glider A towards glider B, which is initially stationary. As glider A approaches glider B, the magnetic poles repel so that B is pushed forward and A moves backward, in the opposite direction to its original motion. The student measures the time for the interrupter card to cross the light gates before and after the interaction. Her results are given below:

Mass of A = 200 g Mass of B = 400 g Length of card = 20.0 cm

Glider	t/s	v/m s^{-1}	p/kg m s^{-1}
A (before)	0.221	0.905	0.181
A (after)	0.659		
B (after)	0.331		

(b) Complete the table to show the velocity and momentum of the gliders after the collision. (2 marks)

(c) Show whether or not these results confirm that momentum is conserved in the interaction. (2 marks)

(d) What is meant by an *elastic* collision? Use the student's results to determine whether the interaction between the gliders is elastic. (3 marks)

Total: 9 marks

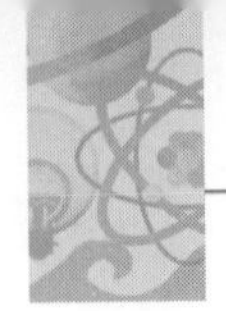

Answer to Question 14

(a) In any system of interacting bodies, the total momentum is conserved ✓ provided that no resultant external force acts on the system. ✓

(b) After the collision:

velocity of A $= \dfrac{-0.20\,\text{m}}{0.659\,\text{s}} = -0.303\,\text{m}\,\text{s}^{-1}$,

velocity of B $= \dfrac{0.20\,\text{m}}{0.331\,\text{s}} = 0.604\,\text{m}\,\text{s}^{-1}$ ✓

momentum of A $= 0.20\,\text{kg} \times (-0.303\,\text{m}\,\text{s}^{-1}) = -0.061\,\text{kg}\,\text{m}\,\text{s}^{-1}$,
momentum of B $= 0.40\,\text{kg} \times 0.604\,\text{m}\,\text{s}^{-1} = 0.242\,\text{kg}\,\text{m}\,\text{s}^{-1}$ ✓

A typical error made by a grade-C candidate is to omit the negative signs for the velocity and momentum of A after the collision; this would lead to one of the marks being lost.

(c) Momentum before the collision $= 0.181\,\text{kg}\,\text{m}\,\text{s}^{-1}$
Total momentum after the collision $= -0.061\,\text{kg}\,\text{m}\,\text{s}^{-1} + 0.242\,\text{kg}\,\text{m}\,\text{s}^{-1}$ ✓
$= 0.181\,\text{kg}\,\text{m}\,\text{s}^{-1}$

so momentum is conserved ✓

(d) An elastic collision is a collision in which the kinetic energy is conserved. ✓

It is essential to refer to *kinetic* energy. Energy by itself is conserved in all collisions (the law of conservation of energy).

E_k before collision $= \frac{1}{2} \times 0.200\,\text{kg} \times (0.905\,\text{m}\,\text{s}^{-1})^2 = 0.082\,\text{J}$

E_k after collision $= \frac{1}{2} \times 0.200\,\text{kg} \times (0.303\,\text{m}\,\text{s}^{-1})^2 + \frac{1}{2} \times 0.400\,\text{kg} \times (0.604\,\text{m}\,\text{s}^{-1})^2$ ✓
$= 0.082\,\text{J}$

hence the collision is elastic ✓

Even if incorrect values of velocity were found in part (a) and used here, it is still possible to gain both marks if the working is correct and the conclusion valid.

Question 15

(a) Outline the alpha particle scattering experiment used by Geiger and Marsden to establish the nuclear model of the atom. You should include details of the results of the experiment and the conclusions drawn from them. (5 marks)
The symbols for an alpha particle and a gold nucleus are given below:

$^{4}_{2}\text{He}$ **alpha particle** $^{197}_{79}\text{Au}$ **gold nucleus**

(b) Explain the meaning of the numbers on the symbols. (2 marks)

(c) Show that the mass of the alpha particle is about 7×10^{-27} kg. (1 mark)

(d) The radius of a gold nucleus is 6.8×10^{-15} m. The alpha particles used for the scattering experiment have sufficient energy to directly approach to within ten times this distance from the nucleus. Calculate the force between a gold nucleus and an alpha particle at this separation. (3 marks)

Total: 11 marks

Answer to Question 15

(a) *Any five of the ticked points below would be accepted*:

- Alpha particles are fired at a (thin) gold film ✓ in a vacuum. ✓
- Most particles pass straight through (or are only slightly deflected). ✓
- A few particles are deviated through large angles (or are reflected straight back). ✓
- These observations suggest that the mass of each gold atom is concentrated in a very small nucleus ✓ which carries a positive charge. ✓

(b) The lower number represents the number of protons in the nucleus (the proton number) and the upper one is the total number of protons and neutrons (the nucleon number). ✓
Thus the alpha particle contains two protons and two neutrons; the gold nucleus contains 79 protons and 118 neutrons. ✓ (*Either one of these two examples would earn the mark.*)

e To gain both marks, reference must be made to at least one of the given symbols.

(c) Mass $\approx 4 \times 1.67 \times 10^{-27}$ kg (neutrons have a similar mass to protons)
$= 6.7 \times 10^{-27}$ kg ✓

(d) $F = \frac{kQ_1Q_2}{r^2}$ ✓

$$= \frac{(9.0 \times 10^{9}\ \text{N m}^2\ \text{C}^{-2}) \times (2 \times 1.6 \times 10^{-19}\ \text{C}) \times (79 \times 1.6 \times 10^{-19}\ \text{C})}{(10 \times 6.8 \times 10^{-15}\ \text{m})^2}$$

$= 7.9$ N ✓

Question 16

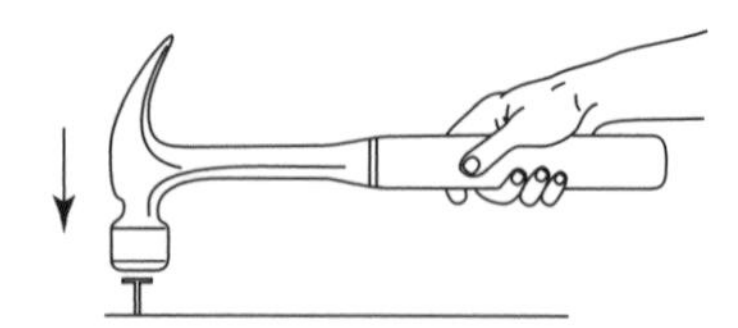

A hammer of mass 0.40 kg is used to drive a nail into a wooden board. The hammer head hits the nail at 6.2 m s^{-1} and rebounds from the nail with a speed of 1.2 m s^{-1}. The hammer is in contact with the nail for 8.5 ms.

(a) Calculate the impulse of the hammer head on the nail. (2 marks)

(b) Calculate the average force exerted by the hammer on the nail. (1 mark)

Total: 3 marks

Answer to Question 16

(a) Taking upwards to be the positive direction,

impulse = change in momentum

$= 0.40\,\text{kg} \times 1.2\,\text{m s}^{-1} - 0.40\,\text{kg} \times (-6.2\,\text{m s}^{-1})$ ✓

$= 3.0\,\text{kg m s}^{-1}$ ✓

(b) Average force $= \dfrac{\text{impulse}}{\text{time}} = \dfrac{3.0\,\text{kg m s}^{-1}}{8.5 \times 10^{-3}\,\text{s}} = 350\,\text{N}$ ✓

Question 17

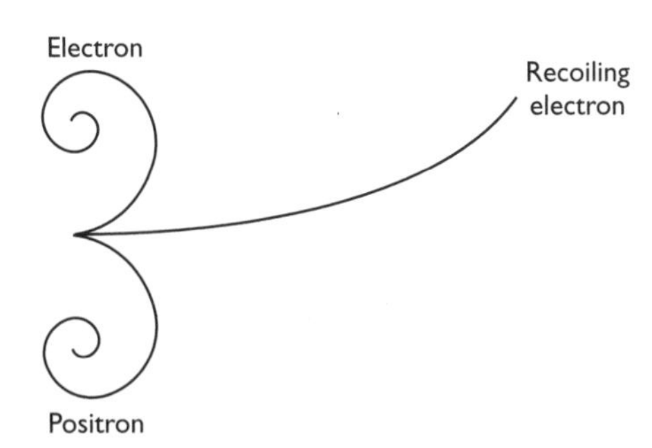

The diagram shows the tracks in a bubble chamber produced when a high-energy gamma ray interacts with a stationary electron to create an electron–positron pair. This reaction is represented by the equation:

$\gamma \rightarrow e^- + e^+$

The third track is that of the originally stationary electron recoiling.

(a) Why is there no trace for the gamma-ray photon? (1 mark)

(b) Explain how the laws of conservation of charge and energy apply to this interaction. (4 marks)

(c) What other conservation law must apply to this event? (1 mark)

(d) The magnetic field across the chamber is perpendicular to the plane of the diagram. Is the direction of the field into or out of the paper? (1 mark)

(e) Explain how you could deduce that the speed of the recoiling electron is greater than the speeds of the electron and positron produced by the reaction. (1 mark)

(f) Calculate the maximum wavelength of the gamma ray needed to create an electron–positron pair, given that the rest mass of an electron is 0.512 MeV/c^2. (4 marks)

Total: 12 marks

Answer to Question 17

(a) The gamma ray is non-ionising (the photons carry no charge), and so does not leave a bubble trail. ✓

(b) Conservation of charge: initial charge carried by γ and e^- = $0 + (-1e) = -1e$
final charge carried by e^-, e^+ and recoiling e^- = $(-1e) + (+1e) + (-1e) = -1e$ ✓

e An answer that compares the initial charge carried by the incident gamma-ray photon (0) and the final charge carried by the electron–positron pair ((−1e) + (+1e) = 0) would also be accepted.

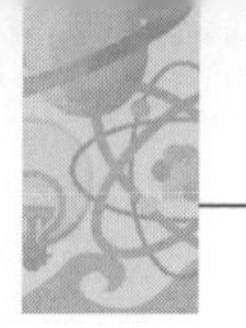

Conservation of energy:
initial energy of photon = hf ✓
final energy = rest mass energy of positron
+ rest mass energy of electron ✓
+ kinetic energy of electron–positron pair and recoil electron ✓

(c) Momentum must also be conserved. ✓

(d) By Fleming's left-hand rule, the magnetic field is out of the paper. ✓

e Conventional current flow is indicated by the direction of the positron.

(e) The path of the recoiling electron has a greater radius than the trails of the electron and positron. ✓ $\left(r = \frac{p}{Bq} = \frac{mv}{Bq}\right)$

(f) Minimum energy needed to create an electron–positron pair is:
$2 \times 0.512\,\text{MeV} = 1.024\,\text{MeV}$ ✓
$= (1.024 \times 10^{6}\,\text{eV}) \times (1.6 \times 10^{-19}\,\text{J eV}^{-1})$ ✓
$= 1.64 \times 10^{-13}\,\text{J}$

Energy of photon $= \frac{hc}{\lambda}$ ✓

$$\lambda_{max} = \frac{hc}{E_{min}} = \frac{6.63 \times 10^{-34}\,\text{J s} \times 3.00 \times 10^{8}\,\text{m s}^{-1}}{1.64 \times 10^{-13}\,\text{J}} = 1.2 \times 10^{-12}\,\text{m}$$ ✓

Question 18

A 100 μF capacitor is charged through a resistor using a battery of emf 6.0 V. The voltage across the capacitor is taken at regular intervals until it is fully charged. The readings for the first minute are given below:

Time/s	0	10.0	20.0	30.0	40.0	50.0	60.0
Voltage/V	0	2.2	3.6	4.5	5.0	5.4	5.6

(a) Plot a graph of voltage against time on the grid provided. (2 marks)

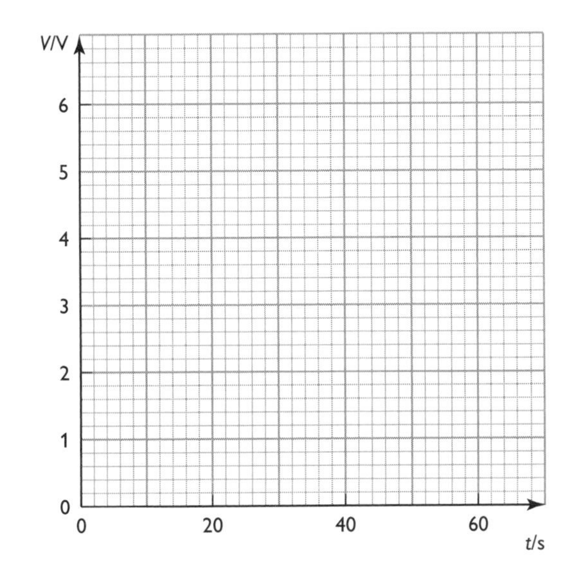

(b) What is meant by the term *time constant* for a capacitor charging through a resistor? (1 mark)

(c) The voltage across the capacitor rises to 63% of the supply voltage in a time equal to one time constant. Use your graph to determine the value of the time constant of the circuit, and so determine the resistance *R*. (3 marks)

(d) Give one practical application of such a circuit. (1 mark)

Total: 7 marks

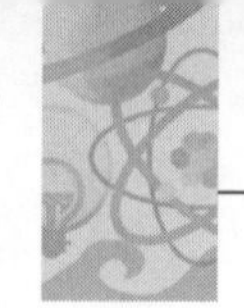

Answer to Question 18

(a)

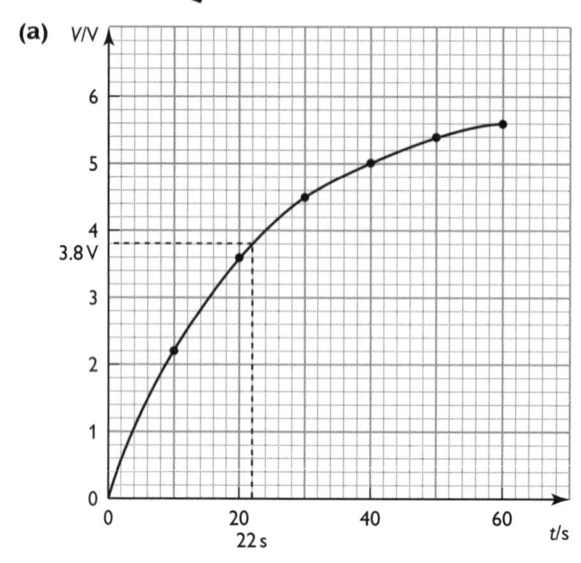

The marks are awarded for:

- data points correctly plotted ✓
- a smooth curve through the plotted points ✓

(b) $T = RC$ ✓

(c) From the graph, $V = 0.63 \times 6.0\,\text{V} = 3.8\,\text{V}$ ✓ when $t = 22\,\text{s}$ ✓

$$R = \frac{T}{C} = \frac{22\,\text{s}}{100 \times 10^{-6}\,\text{F}} = 2.2 \times 10^{5}\,\Omega = 220\,\text{k}\Omega$$ ✓

(d) *Any timing device.* ✓

Question 19

A petrol engine ignition system uses an induction coil to generate the high voltage needed for creating a spark across the electrodes of a spark plug to ignite the fuel–air mixture in the cylinders. A schematic diagram is shown below:

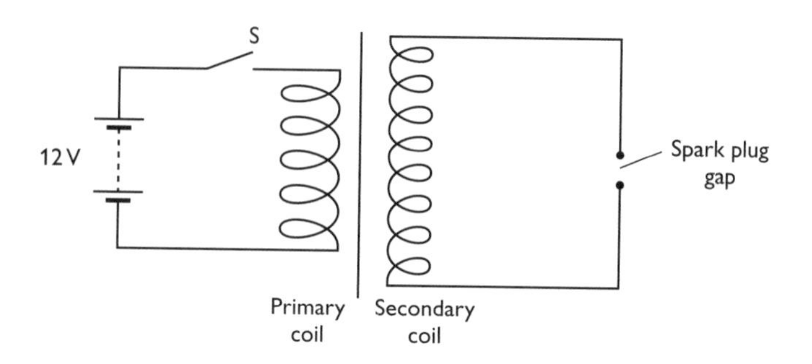

(a) For a spark to jump between the electrodes, the electric field strength needs to be $3.0 \times 10^7\,\text{V m}^{-1}$. Show that, for a gap of 0.7 mm, the voltage across the plug is about 20 kV. (2 marks)

To produce such high voltages, the primary circuit is broken so that the current falls rapidly to zero.

(b) Use Faraday's law to explain how this leads to a high voltage being induced in the secondary coil. (4 marks)

(c) The secondary coil in one system has 20 000 turns and a cross-sectional area of $4.0 \times 10^{-3}\,\text{m}^2$. The flux density linked with the coils is 1.2 T, which drops to zero in 3.6 ms when the circuit of the primary coil is broken.
Calculate:
(i) the flux linkage of the secondary coil before the circuit is broken (1 mark)
(ii) the voltage induced across the ends of the secondary coil (2 marks)

Total: 9 marks

■ ■ ■

Answer to Question 19

(a) $E = \frac{V}{d}$ ✓ $\Rightarrow V = Ed = (3.0 \times 10^7\,\text{V m}^{-1}) \times (0.7 \times 10^{-3}\,\text{m}) = 2.1 \times 10^4\,\text{V} = 21\,\text{kV}$ ✓

(b) Faraday's law states that the induced emf is proportional to the rate of change of flux linkage in a conductor. ✓

When the primary circuit is broken, the magnetic flux due to the current in the primary coil falls rapidly to zero. ✓ This causes a rapid change in the flux linkage with the secondary coil, ✓ which has a large number of turns, ✓ and so a large emf is induced in the secondary coil.

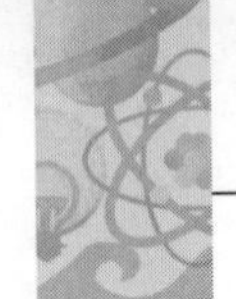

The question specifically asks for the use of Faraday's law, so you cannot achieve full marks unless you include a statement of the law in your answer.

(c) (i) Flux linkage = $N\Phi = NBA = (2.0 \times 10^4) \times 1.2\,\text{T} \times (4.0 \times 10^{-3}\,\text{m}^2) = 96\,\text{Wb}$ ✓

(ii) $\varepsilon = -\dfrac{\text{d}(N\Phi)}{\text{d}t}$ ✓ $= -\dfrac{(96\,\text{Wb} - 0\,\text{Wb})}{3.6 \times 10^{-3}\,\text{s}} = -27\,\text{kV}$ ✓

Test Paper 2

Questions 1–10

For questions 1–10, select one answer from A to D.

(1) The nucleus of a nitrogen atom is represented by the symbol $^{14}_{7}N$. The nucleus consists of:

A 7 neutrons and 14 protons
B 7 protons and 7 electrons
C 7 protons and 7 neutrons
D 7 protons and 14 neutrons

(1 mark)

(2) When the nitrogen nucleus in question 1 is accelerated through a potential difference of 4000 V, the kinetic energy it gains is:

A 28 keV
B 56 keV
C 28 MeV
D 56 MeV

(1 mark)

(3) The angular speed of the Moon around the Earth is approximately:

A $2\pi \times 10^{-7}$ rad s^{-1}
B $4\pi \times 10^{-7}$ rad s^{-1}
C $8\pi \times 10^{-7}$ rad s^{-1}
D $8\pi \times 10^{6}$ rad s^{-1}

(1 mark)

(4) Which of the following will *not* experience a force?

A an electron moving parallel to an electric field
B an electron moving parallel to a magnetic field
C an electron moving perpendicular to an electric field
D a stationary electron in an electric field

(1 mark)

Questions 5 and 6 relate to the nuclear interaction represented by the equation $^{7}_{3}Li + ^{1}_{1}H \rightarrow 2X$

(5) X represents:

A an alpha particle
B a beta particle
C a gamma ray
D a positron

(1 mark)

(6) The *mass defect* Δm (the difference in mass between the constituents on each side of the equation) is 3.1×10^{-29} kg. The energy released in this process is:

A 4.7×10^{-21} J
B 9.3×10^{-21} J
C 1.4×10^{-12} J
D 2.8×10^{-12} J

(1 mark)

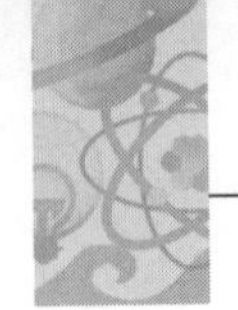

The following are four possible graphs of a quantity *Y* plotted against another quantity *X*. Refer to these graphs when answering questions 7, 8 and 9.

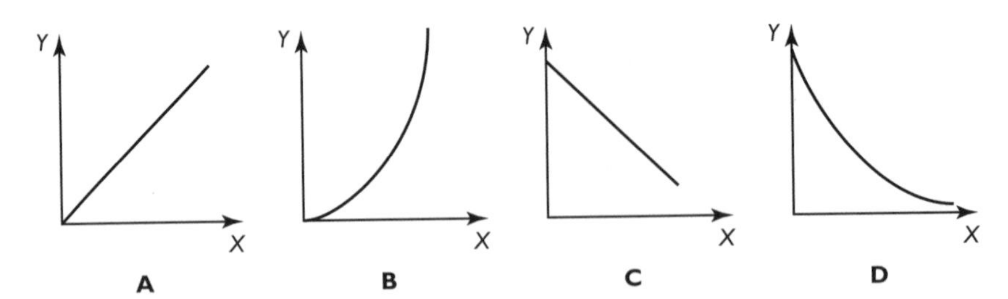

(7) Which graph *best* represents *Y* when it is the force acting on a wire carrying a steady current at right angles to a uniform magnetic field and *X* is the length of wire within the field? (1 mark)

(8) Which graph *best* represents *Y* when it is the kinetic energy of a proton and *X* is the proton's momentum? (1 mark)

(9) Which graph *best* represents *Y* when it is the natural logarithm of the charge stored on a capacitor (ln Q) and *X* is the time that the capacitor has been discharging through a resistor? (1 mark)

(10) A π^- meson is composed of which combination of quarks?

A ud

B $\bar{u}d$

C $u\bar{d}$

D $\bar{u}\bar{d}$

(1 mark)

Total: 10 marks

Answers to Questions 1–10

(1) C

The lower number is the proton number, and the upper number is the nucleon number — the *total* number of protons and neutrons. There are no electrons in the nucleus.

(2) A

The charge on the nitrogen nucleus is +7*e*, where e represents the magnitude of charge on an electron. Therefore, from $W = VQ$, the work done on the nucleus is 4000×7 electronvolts, i.e. 28 keV.

(3) C

This comes from $\omega = \frac{2\pi}{T}$, where *T* is the period of the Moon's orbit, $T = 28 \times 24 \times 60 \times 60\,\text{s} = 2.4 \times 10^6\,\text{s}$.

(4) B

In an electric field, any charged particle will experience a force, whatever its motion. In a magnetic field, however, the charge needs to be moving, with some component of its velocity perpendicular to the field, in order to experience a force.

(5) A

For the equation to balance, X must have a proton number of 2 and a nucleon number of 4. An alpha particle $\left({}^{4}_{2}\text{He}\right)$ satisfies these conditions.

(6) D

Use $\Delta E = c^2 \Delta m = (3.0 \times 10^8\,\text{m s}^{-1})^2 \times (3.1 \times 10^{-29}\,\text{kg})$.

(7) A

The current is at right angles to the field, so $F = BIl$. Since B and I are constant, F is proportional to l.

(8) B

$E_k = p^2/2m$, so Y is proportional to X^2 and the graph of Y against X is a parabola.

(9) C

$Q = Q_0 e^{-t/RC}$ so $\ln Q = \ln Q_0 - \frac{t}{RC}$, so the graph is a straight line with negative gradient.

(10) B

A meson must consist of a quark–antiquark pair, so choices A and D are excluded. For π^- the charge is −1e, and this fits with choice B: anti-up has a charge of $-\frac{2}{3}$e and the down quark has a charge of $-\frac{1}{3}$e.

Question 11

A bar magnet is dropped through a coil of copper wire which is connected to the input of a data-logging device. The data logger displays a voltage–time graph as shown below:

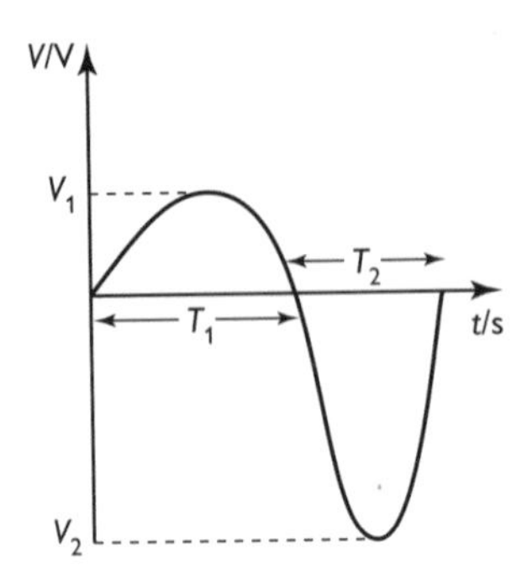

Explain the shape of the graph, making reference to the relative values of V_1 and V_2 and of T_1 and T_2.

Total: 3 marks

Answer to Question 11

When the magnet leaves the coil, it is moving faster than it was when entering the coil, so the induced emf is greater, i.e. the magnitude of V_2 is bigger than the magnitude of V_1. ✓

As the magnet leaves the coil, the direction of the emf is reversed, so V_1 and V_2 have opposite signs. ✓

The magnet accelerates as it drops through the coil, so it takes less time to exit than to enter; thus the duration T_2 will be shorter than the duration T_1. ✓

e A typical grade-C candidate may adequately describe the induction process but fail to specifically compare the values or polarities of the *V*s and *T*s.

Question 12

The table includes four fundamental particles. For each particle, indicate its nature by ticking the appropriate column or columns.

Particle	Baryon	Hadron	Lepton
Neutron			
Positron			
π^+ meson			
Tau neutrino			

Total: 4 marks

Answer to Question 12

neutron: *both* baryon *and* hadron ✓

positron: lepton *only* ✓

π^+ meson: hadron *only* ✓

tau neutrino: lepton *only* ✓

Question 13

In an experiment to investigate the charge on an electron, Robert Millikan observed charged oil droplets in the electric field between a pair of charged parallel plates.

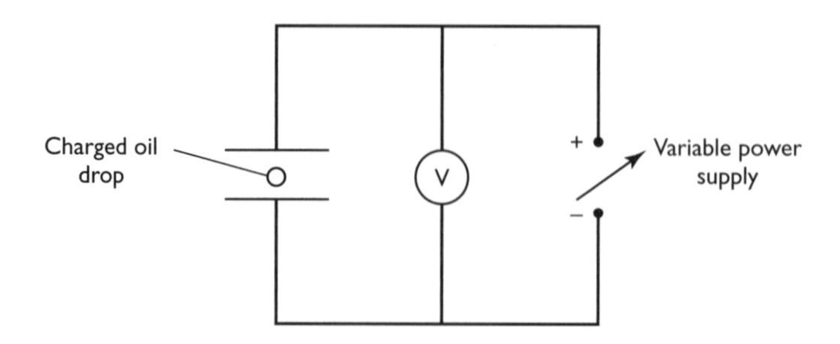

(a) Draw a diagram to show the pattern of the electric field between the plates. (2 marks)

A droplet is seen moving upwards between the plates. The voltage is adjusted so that the droplet stops moving and is held stationary in the field.

(b) Draw a free-body diagram for a charged droplet within the field. Label all the forces acting on the droplet. (2 marks)

Given that the mass of the droplet is 9.8 × 10^{-16} kg, the separation of the plates is 0.80 cm and the potential difference across the plates needed to hold the droplet stationary in the field is 240 V, calculate:

(c) the electric field strength between the plates (2 marks)

(d) the magnitude of the charge on the oil drop (3 marks)

Total: 9 marks

■ ■ ■

Answer to Question 13

(a) The diagram should have:

- lines drawn between the parallel plates, at right angles to the plates ✓
- equal spacing between consecutive lines *and* arrows on the lines pointing from positive to negative ✓

Many candidates lose marks by drawing irregularly spaced lines or by omitting the direction. It is important to remember that a uniform field must be represented by evenly spaced parallel lines.

(b) The diagram should include:

- an upward force labelled 'electric force' or Eq ✓
- a downward force labelled 'weight', W or mg ✓

(c) $E = \frac{V}{d} = \frac{240\,\text{V}}{0.8 \times 10^{-2}\,\text{m}}$ ✓ $= 3.0 \times 10^{4}\,\text{V m}^{-1}$ ✓

(d) $Eq = mg$ ✓ so $q = \frac{mg}{E} = \frac{9.8 \times 10^{-16}\,\text{kg} \times 9.8\,\text{m s}^{-2}}{3.0 \times 10^{4}\,\text{N C}^{-1}}$ ✓ $= 3.2 \times 10^{-19}\,\text{C}$ ✓

Question 14

A car pulls out of a junction without seeing an oncoming vehicle. The driver turns and then brakes to a halt, but is unable to avoid an oblique collision with the other car. The moving car has a mass of 1500 kg and is travelling at 10 m s^{-1} at the instant the collision occurs. It deflects off the side of the stationary vehicle at an angle of 20° to its initial direction. The stationary car, of mass 2000 kg, is shunted with an initial velocity of 4.0 m s^{-1} at 60° in the other direction, as shown in the diagram.

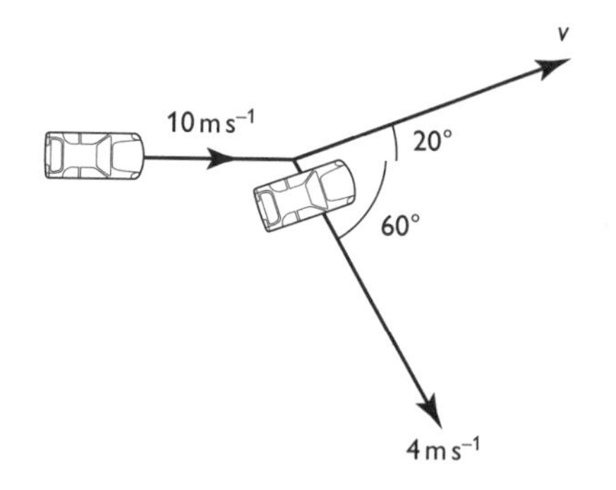

(a) If the velocity of the deflected car is *v*, write down the components, in the original direction of the moving car, of the linear momentum of the two cars immediately after the impact. (2 marks)

(b) Show that *v* is about 8 m s^{-1}. (2 marks)

(c) If the duration of the impact is 0.25 s, calculate the average force on the side of the stationary car. (2 marks)

(d) Modern cars have 'side impact protection systems' (SIPS) that allow the doors to crumple on impact. Explain how this system can reduce the effect of a side collision on the occupants of a car involved in an accident. (2 marks)

Total: 8 marks

Answer to Question 14

(a) The components are 1500 kg × $v\cos 20°$ ✓ and 2000 kg × 4.0 m s^{-1} × cos 60° ✓

(b) Using the principle of conservation of linear momentum:

1500 kg × 10 m s^{-1} + 0 kg m s^{-1} = 1500 kg × $v\cos 20°$ + 2000 kg × 4.0 m s^{-1} × cos 60° ✓

which gives v = 7.8 m s^{-1} ✓

(c) Average force $= \dfrac{\text{change in momentum}}{\text{time}}$ in the direction the car is shunted ✓

$$= \frac{2000\,\text{kg} \times 4.0\,\text{m s}^{-1} - 0\,\text{kg m s}^{-1}}{0.25\,\text{s}} = 32\,\text{kN} ✓$$

A grade-A candidate will realise that the force of impact on the stationary car must be in the direction the car is shunted. A grade-C candidate may use the component in the initial direction of the moving car, and hence gain just one of the marks.

(d) The crumple zone extends the duration of the impact ✓ and so reduces the resultant force on the car. ✓

Question 15

A capacitor is charged using a 12 V battery and then discharged through a 220 kΩ resistor. The discharge current is measured at regular intervals and a graph of current against time is plotted.

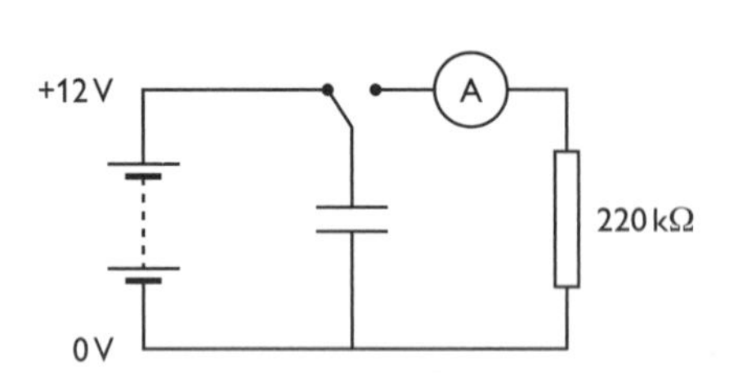

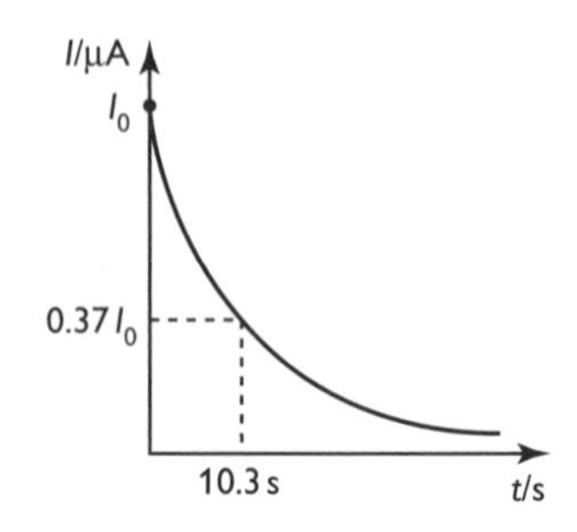

(a) Determine the initial value of the current, I_0. (1 mark)

(b) How could you use the graph to estimate the initial charge stored on the capacitor? (1 mark)

(c) The time taken for the current to fall to 37% of the initial value is equal to one time constant, and is found to be 10.3 s. Define the term *time constant* for a capacitor discharging through a resistor, and calculate the value of the capacitance of the capacitor. (2 marks)

(d) Calculate the value of *I* after 20.6 s. (2 marks)

The resistor in the circuit is replaced by an electric motor, and a 10 000 μF capacitor is used in place of the original one. When the capacitor is fully charged and then discharged through the motor, a load of 20 g is raised through a height of 80 cm by the motor.

(e) Calculate the energy stored in the 10 000 μF capacitor when it is fully charged. (2 marks)

(f) Calculate the work done by the motor in raising the load, and determine the efficiency of the system. (3 marks)

Total: 11 marks

Answer to Question 15

(a) $I_0 = \frac{V_0}{R} = \frac{12\,\text{V}}{220 \times 10^3\,\Omega} = 5.5 \times 10^{-5}\,\text{A} = 55\,\mu\text{A}$ ✓

(b) Q = area under the I–t graph ✓

(c) The time constant for the circuit is RC ✓

Here we have $220 \times 10^3 \Omega \times C = 10.3\,\text{s}$, so $C = \dfrac{10.3\,\text{s}}{220 \times 10^3\,\Omega} = 4.7 \times 10^{-5}\,\text{F} = 47\,\mu\text{F}$ ✓

(d) 20.6s is equal to two time constants. ✓

In 10.3s, I falls to $0.37 \times 55\,\mu\text{A}$ and in the next 10.3s it falls to $0.37 \times (0.37 \times 55\,\mu\text{A}) = 7.5\,\mu\text{A}$ ✓

An equally valid method is to substitute $t = 20.6\,\text{s}$ into the formula $I = I_0 e^{-t/RC}$; this would give the answer as 7.4 μA.

(e) $E = \frac{1}{2}CV^2 = \frac{1}{2} \times (10000 \times 10^{-6}\,\text{F}) \times (12\,\text{V})^2$ ✓ $= 0.72\,\text{J}$ ✓

(f) $W = mgh = 0.020\,\text{kg} \times 9.8\,\text{m}\,\text{s}^{-2} \times 0.80\,\text{m} = 0.16\,\text{J}$ ✓

Efficiency $= \dfrac{0.16\,\text{J}}{0.72\,\text{J}} \times 100\%$ ✓ $= 22\%$ ✓

Question 16

(a) Explain why a body moving at constant speed in a circular path needs a resultant force acting on it. (2 marks)

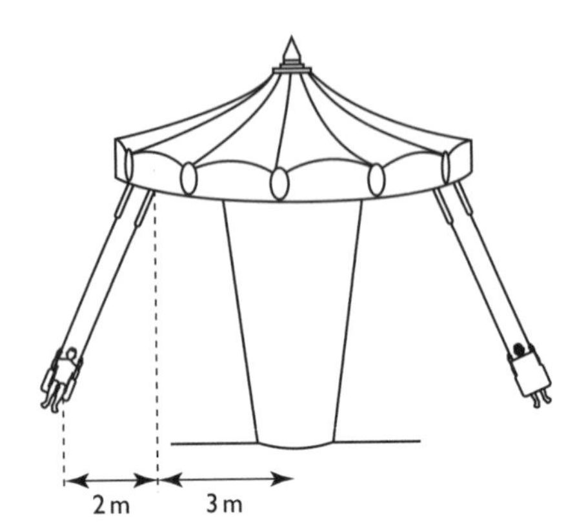

A carousel has a central section of radius 3.0 m with chairs attached to the perimeter on chains that are 4.0 m long. When the carousel rotates, the chairs, with their occupants, move out a further 2.0 m so that the chains make an angle of 30° to the vertical and the riders move in a horizontal circle at constant speed.

(b) Draw a free-body force diagram for a chair and its rider. (2 marks)

(c) If the mass of the chair plus its occupant is 120 kg, calculate the tension in the chain. (2 marks)

(d) Determine the speed of the rider around the circular path. (3 marks)

Total: 9 marks

Answer to Question 16

(a) The *direction* of the (tangential) velocity is changing, ✓ so there must be an *acceleration*, which needs a *force* to produce it. ✓

(b) The diagram should include:

- a force labelled 'weight', W or mg acting downwards ✓
- a force labelled 'tension', T or F acting along the chain ✓

e Only these two forces should be drawn. Some candidates draw components of the tension (e.g. vertically up and horizontally towards the central axis of the carousel) onto the free-body force diagram, but this causes them to lose a mark.

(c) The chair and rider are in equilibrium in the vertical plane, so:

$mg = T\cos 30°$

$T = \dfrac{120\,\text{kg} \times 9.81\,\text{m s}^{-2}}{\cos 30°}$ ✓ $= 1360\,\text{N}$ ✓

(d) The centripetal force is provided by the horizontal component of the tension, so:

$\dfrac{mv^2}{r} = T\sin 30°$ ✓

$\dfrac{120\,\text{kg} \times v^2}{3.0\,\text{m} + 2.0\,\text{m}} = 1360\,\text{N} \times \sin 30° = 680\,\text{N}$ ✓

hence $v = 5.3\,\text{m s}^{-1}$ ✓

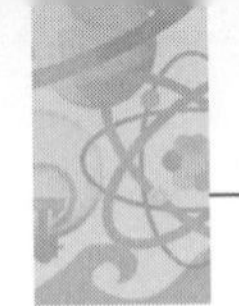

Question 17

Neutrons, like electrons and X-rays, can be used to examine crystal structure. A suitable de Broglie wavelength for the neutrons is about 1 nm.

(a) Explain why neutrons must have a de Broglie wavelength of this order of magnitude. (2 marks)

(b) Give one advantage and one disadvantage of using neutrons rather than protons for this purpose. (2 marks)

(c) Given that the mass of a neutron is 1.67×10^{-27} kg, calculate the kinetic energy of a neutron that has a de Broglie wavelength of 1.10 nm. (3 marks)

(d) What is meant by *wave–particle duality*? Illustrate your answer with the example of the neutrons. (3 marks)

Total: 10 marks

Answer to Question 17

(a) The neutrons must have a de Broglie wavelength of about 1 nm because this is the order of magnitude of the spacing of atoms in a crystal. ✓ Having a wavelength that is similar to the size of the 'gap' between atoms is a necessary condition for diffraction to take place. ✓

(b) One advantage of using neutrons is that they are uncharged and so will not experience the repulsive forces from ions within the crystal that protons, which carry a positive charge, would encounter. ✓

One disadvantage is that compared with protons, the momentum, and hence the wavelength, of neutrons is much more difficult to control, because neutrons cannot be accelerated using electric or magnetic fields. ✓

e A grade-C candidate may comment on the charge carried by a proton without saying why this can be an advantage or disadvantage.

(c) $\lambda = \dfrac{h}{p} \Rightarrow p = \dfrac{h}{\lambda} = \dfrac{6.63 \times 10^{-34}\ \text{J s}}{1.10 \times 10^{-9}\ \text{m}}$ ✓ $= 6.03 \times 10^{-25}\ \text{kg m s}^{-1}$

$E_\text{k} = \dfrac{p^2}{2m} = \dfrac{(6.03 \times 10^{-25}\ \text{kg m s}^{-1})^2}{2 \times (1.67 \times 10^{-27}\ \text{kg})}$ ✓ $= 1.1 \times 10^{-22}\ \text{J}$ ✓

(d) Wave–particle duality is when something can act as a particle or a wave depending on the situation. ✓

For example, a neutron behaves like a particle when it is in the nucleus of an atom or when it is used to bombard nuclei in reactors. ✓

But neutrons can also behave like waves — for instance, when they are diffracted by crystal layers. ✓

e The response of a grade-C candidate may include an explanation of wave–particle duality but omit reference to *neutrons*, which was specifically asked for in the question.

Question 18

(a) State Lenz's law of electromagnetic induction. (2 marks)

Some trains and large vehicles use electromagnetic braking systems. A metal disc, or driveshaft, rotating in a magnetic field generates 'eddy currents' that create a retarding force on the disc.

A student investigating electromagnetic braking used a glider on an air track as shown in the diagram below:

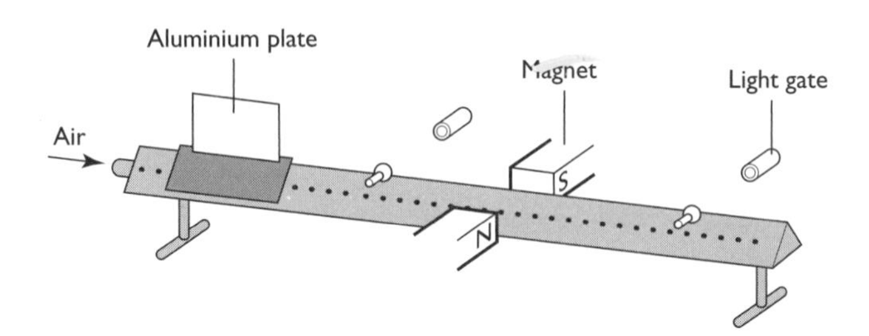

The glider was launched, with a range of speeds, so that an aluminium plate carried by it would pass between the poles of an electromagnet. The plate also acted as an interrupter card so that the speed of the glider could be calculated before and after its passage through the field. The acceleration (proportional to the retarding force) was calculated using the equation:

$$v^2 = u^2 + 2as$$

where u is the speed of the glider before it passed through the field, v is its speed after it had passed between the poles and s is the length of the aluminium plate.

Some of the student's readings are shown in the following table:

u/m s^{-1}	v/m s^{-1}	s/m	a/m s^{-2}
0.41	0.37	0.200	
0.79	0.75	0.200	
1.20	1.16	0.200	
1.62	1.58	0.200	
1.99	1.95	0.200	

(b) Complete the table by calculating the acceleration for each velocity. (2 marks)

(c) Plot a graph of acceleration against initial velocity. (3 marks)

(d) With reference to the graph, use the laws of electromagnetic induction to explain why trains using this form of braking still need to apply conventional friction brakes as they slow down to a halt. (4 marks)

(e) Electric and hybrid cars use *regenerative braking*. At higher speeds, the car can be slowed down by reversing the direction of the electric motor so that it acts as a generator.

(i) Use the principle of conservation of energy to explain how the system operates. (3 marks)

(ii) Give two advantages of regenerative braking over conventional systems. (2 marks)

Total: 16 marks

■ ■ ■

Answer to Question 18

(a) Lenz's law states that the induced current ✓ flows in such a direction as to oppose the change producing it. ✓

Some candidates lose a mark by referring to the induced emf rather than the induced current. It is essential that a *current* flows for the conservation of energy argument (which underlies Lenz's law) to apply.

(b) The values of $a/\mathrm{m\,s^{-2}}$ in the table are, from top to bottom:
–0.078, –0.154, –0.236, –0.320, –0.394

Any four correct magnitudes gain 1 mark (two significant figures are acceptable). ✓ The second mark is for the negative signs. ✓

(c) The marks for the graph are allocated as follows:

- axes correctly labelled, with units ✓
- sensible scale — at least half of the grid should be utilised ✓
- straight line drawn ✓ (either positive or negative slope will be accepted)

(d) The graph shows ✓ that the acceleration (or deceleration, or retarding force) gets weaker as the glider slows down. ✓
This is because, when the glider is moving faster, the induced currents (eddy currents) are larger ✓ and the rate of change of flux linkage is greater. ✓

A grade-A candidate would be expected to refer to the graph and link the retarding force to the speed and the induced current.

(e) (i) In order that an electric current can be generated, energy must be transferred from another form. ✓ In this case, the kinetic energy lost by the car is converted into electrical energy to recharge the battery ✓ (*or: the kinetic energy converted into chemical energy*). By Lenz's law, the induced current in the generator opposes the rotation ✓ and acts as a brake.

(ii) Some of the kinetic energy lost in braking is recovered in another form. ✓ There is less wear on the frictional brake pads. ✓